Application of Artificial Intelligence in New Materials Discovery

Edited by

Inamuddin[1], Maha Khan[1], Mohammad A. Jafar Mazumder[2,3],

[1]Department of Applied Chemistry, Zakir Husain College of Engineering and Technology, Faculty of Engineering and Technology, Aligarh Muslim University, Aligarh-202002, India

[2]Chemistry Department, King Fahd University of Petroleum & Minerals, Dhahran 31261, Saudi Arabia

[3]Interdisciplinary Research Center for Advanced Materials, King Fahd University of Petroleum & Minerals, Dhahran 31261, Saudi Arabia

Copyright © 2023 by the authors

Published by **Materials Research Forum LLC**
Millersville, PA 17551, USA

All rights reserved. No part of the contents of this book may be reproduced or transmitted in any form or by any means without the written permission of the publisher.

Published as part of the book series
Materials Research Foundations
Volume 147 (2023)
ISSN 2471-8890 (Print)
ISSN 2471-8904 (Online)

Print ISBN 978-1-64490-252-3
eBook ISBN 978-1-64490-253-0

This book contains information obtained from authentic and highly regarded sources. Reasonable efforts have been made to publish reliable data and information, but the author and publisher cannot assume responsibility for the validity of all materials or the consequences of their use. The authors and publishers have attempted to trace the copyright holders of all material reproduced in this publication and apologize to copyright holders if permission to publish in this form has not been obtained. If any copyright material has not been acknowledged please write and let us know so we may rectify this in any future reprints.

Distributed worldwide by

Materials Research Forum LLC
105 Springdale Lane
Millersville, PA 17551
USA
https://www.mrforum.com

Manufactured in the United States of America
10 9 8 7 6 5 4 3 2 1

Table of Contents

Preface

Artificial Intelligence Nano-Robots
Eksha Guliani, Anita Gupta, Tejendra K Gupta 1

Data Mining in Material Science
Moganapriya Chinnasamy, Rajasekar Rathanasamy, Samir Kumar Pal,
Manoj Kumar Kathiresan, Sathish Kumar Palaniappan 24

Artificial Intelligence Applications in Solar Photovoltaic Renewable Energy Systems
Ifeanyi Michael Smarte Anekwe, Emmanuel Kweinor Tetteh,
Edward Kwaku Armah... 47

Artificial Intelligence in Material Genomics
Joy Hoskeri H, Nivedita Pujari S, Badrinath Kulkarni, Arun K. Shettar 87

Applications of Artificial Intelligence in Polymer Manufacturing
Satyansh Srivastava, Bhoomika Varshney, V.P. Sharma, Babra Ali3 105

Artificial Intelligence for Energy Conversion
Tapasi Ghosh, Bhargavi Koneru, Prasun Banerjee....................... 123

Keyword Index
About the Editors

Preface

Researchers in the field of materials science have been attempting to develop innovative materials with enhanced operating qualities while still fulfilling economic and environmental standards for a long time. The key purpose of this activity is to produce materials with outstanding attributes such as hardness for tools, heat resistance for insulators, wear or abrasion resistance, corrosion resistance, elasticity for flexible systems, ecofriendly, easy recycling, cheap cost, and other characteristics. Meeting these criteria has become even more difficult as the global free-market economy continues to evolve at a rapid pace. Engineered materials, which arose as an outcome of scientific progress near the close of the twentieth century, are currently discovering new ways to be produced and shaped. At the moment, there is a continuous trend away from empirical methods of examining reality in favor of new and advanced technology, mostly computer-based procedures that start with mathematical models of the subject of research.

These techniques including artificial neural networks, Bayesian, support vector machines, heuristics, etc. have been successfully applied in the prediction of crystal structure, component prediction, process optimization, finding solutions to DFT (density functional theory), monitoring, or classification problems in material science.

Material science that once was considered a branch of art gradually shifted towards science on mathematical application and now has become the ultimate topic of artificial intelligence. With the popularity of Google DeepMind and the success of the alpha-zero algorithm in solving problems related to protein unfolding, there is a growing interest amongst researchers to learn, apply and include artificial intelligence as an integral part of their research just like mathematic and allied techniques.

This book examines the function and use of artificial intelligence in solving complicated and time-consuming issues in material science in order to obtain improved and desired qualities for the material under investigation.

Chapter 1 discusses the role of robots and artificial intelligence in practical service solutions, and their importance to understand technology in value co-creation. In this chapter, the role of composite materials, shape memory polymers, and liquid crystals in robotic devices are discussed. The next section of this chapter explains the components and materials, movements in nanorobots, and mechanism. Finally, the future challenges and concluding remarks along with future scope are addressed.

Chapter 2 briefly summaries the evolution of materials science research, followed by an emphasis on the key principles and basic processes of AI technique. Machine learning applications in materials science include novel material discovery, material property predicting and other objectives spanning from macro to micro level are discussed in detail.

Chapter 3 provides a review of the application of artificial intelligence in a solar PV system and highlights the challenges and prospects for its effective implementation in a renewable energy system. In addition, an overview of the application of artificial intelligence in renewable energy systems was presented.

Chapter 4 showcases some of the key advancements in the field of artificial intelligence and its reflections in the discovery and development of material genomics. This chapter also covers generation-wise growth in the artificial intelligence domain and its subsequent implications in high throughput material synthesis and fabrication.

Chapter 5 discusses that artificial intelligence (AI) is multidisciplinary with extraordinary potential and opportunities for the development of innovative polymers. It involves computational, experimental, theoretical, and state of art approaches through machine learning, data-driven modes as well as mechanistic insights under regulatory compliance.

Chapter 6 demonstrates the advancement of machine learning and deep learning techniques for the exploration of the resources and methodologies for energy conversion. The major focus is given on how the power of AI can be utilized to facilitate the process of discovering new materials as alternative energy sources.

Editors

Chapter 1

Artificial Intelligence Nano-Robots

Eksha Guliani[1], Anita Gupta[1] and Tejendra K Gupta[1*]

[1]Amity Institute of Applied Sciences, Amity University, Sector-125, Noida 201313, India

*tkgupta@amity.edu

Abstract

The interaction of humans with technology is greatly connected and built up by trust, it is an inevitable user acceptance that is required at every step. There is an existence of applications of artificial intelligence (AI) and robotics in the field of engineering, architecture, and construction industry. Therefore, there is a requirement for carrying out studies related to the trust aspect in every system. As artificial intelligence (AI) and robots are increasingly taking place in practical service solutions, it is necessary to understand technology in value co-creation. There has been a tremendous advancement in the field of nanotechnology which has ultimately led to the production of nanorobots, and these have a wide range of applications in nanomedicine. Thus, a nanorobot being an artificial device should be employed and used as a strategy to escape from such kind of immune system. To prevent barriers during the movement trajectory, a self-organized trajectory should be planned for a nanorobot to function accordingly. Due to some of the practical limitations, all the scientists could afford to compromise the product design so that it suits the needs in various stages from a unique design position. This revolution might place the twenty-first century at a decisive position in the era of history.

Keywords

Robotics, Nanorobots, Nanotechnology, Materials, Composites

Contents

Artificial Intelligence Nano-Robots 1

1. Introduction 3
2. Composites 5

	2.1	Liquid crystal elastomers 5
	2.2	Shape memory polymers 6
	2.3	Hydrogels 7
	2.4	CNT actuators 7
	2.5	Conducting polymers 8
3.	Components and materials 9
4.	Movement in nanorobots 10
5.	Mechanism and stimulation 11
6.	Trust dimensions 14
	6.1	Reliability and safety 14
	6.2	Explainability and interpretability 14
	6.3	Privacy and security 14
	6.4	Performance and robustness 15
7.	Actuators 15
	7.1	Thermally responsive actuators 15
	7.2	Photo-responsive actuators 16
	7.3	Magnetically responsive actuators 16
	7.4	Electrically responsive actuators 16
8.	Applications 17
	8.1	Cancer detection and its treatment 17
	8.2	Nanorobots in the diagnosis and treatment of diabetes 17
	8.3	Artificial oxygen carrier nanorobot 17
9.	Future challenges 17

Conclusion and future scope 18

Conflict of interest 18

Acknowledgment 18

References 18

1. Introduction

Technology has now made an important place in everyone's life. In today's scenario, due to the burgeoning population, there is a great advancement in the field of science and technology. Smartphones nowadays wake up from sleep and there are automatic toasters that can prepare breakfast while in bed. These are the machines that mimic the cognitive functions of human beings and can be easily programmed to perform complex actions automatically with great ease. These machines/gadgets are hence known as artificially intelligent machines which are commonly called robots. They can access a basic daily routine by extracting information from various calendars, tracking the sleep cycle, and start preparing meals by knowing the right times from daily routines. This has become possible only because of artificial intelligence which makes robots common in markets, workplaces, restaurants, etc., and therefore, robots can change the lifestyle of people and emerge as the machines which are the future of human beings [1-10]. Just imagine that communication to the workplace in a robotic driverless car would give us time to read, receive calls, or even take a nap. The robotic car can even serve as our assistant which can keep a record of all the chores that people do and plan the best route that must be followed to reach the destination well on time by checking the traffic data and eliminating the congested route [11]. Robots can very well assist humans in their tasks. In this direction, tremendous advancements made over the decades have shown that robotic devices can manipulate, interact and locomote with the surroundings in different ways. They can recognize objects, learn to improve control, acquire new knowledge, coordinate a team, and can imitate simple motions performed by humans [12-15].

The latest field of technology is nanotechnology which best describes the activities which are considered mainly at the very basic starting from atomic and molecular level as these have tremendous applications in the real world. So, to achieve cost-effectiveness in the field of nanotechnology, it is necessary to cause molecular manufacturing automatically. The process of manufacturing or the engineering related to the products at the molecular level must be processed through robotic devices that are termed nano-robots [16-18]. The interaction of mankind and technology is very much connected with the trust which is an inevitable requirement of acceptance by the user. Among several applications of artificial intelligence, robotics has emerged in every sphere of life including architecture, construction, and engineering. Robotic work on artificial intelligence is nowadays essential in organizational and industrial processes [19-21]. It has become prominent in our lives and with the emergence of intelligent agents in an industry, there is an ability to the formation of trust that has become more apparent in today's scenario. There are certain limitations that scientists and engineers have encountered while designing the product data required to fit at different stages from a traditional design to the improvised version. The design required for traditional manufacturing causes a reduction in the potential efficiency but when it is combined with artificial technology, it boosts the industries like health care, sensing, robotics, aerospace, etc. which make nanorobotics an essential branch in the field of science and technology [22-30].

To achieve the pace of robotics, it is essential to go beyond them with faster development of easily adaptable technologies. One way is to relate with the nano-robots this thinking of the artificial organisms. Conventional robots can be easily divided into electrical, computational, and electrical domains and artificial organisms or nano-robots can be thought of as having several components in their system which act as the body. Therefore, it can be said that functional equivalence is there that acts along with the natural organism through a robotic organism, called the brain and it is very much similar to the control system and even acts as a computer. The body of the natural organism resembles the mechanical structure of the nanorobot, and the stomach of the natural organism is the power of the robot just like a battery or a cell [31]. A related diagram is shown in Figure 1.

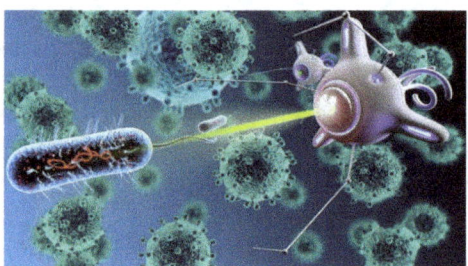

Figure 1: Illustration of a nanorobot targeting the active site. Adopted from Big Data Analytic News website (https://voi.id/en/technology/86234/3-tiny-robots-that-changed-the-world-some-can-break-down-plastic).

A play was performed by Karel Capek in 1920, which used the word robot for the very first time. Then in 1932, Japan was the first country to introduce a toy robot called Lilliput that had a rigid, canonical, and stocky appearance. It was about 15 cm tall that could walk with a thin figure. After that, in 1942, Isaac Asimov showed three laws required for the working of robots. In the year 1950, Alan Turning formulated a test drive that could measure the artificial intelligence of computers. The first and foremost programmable robot named Unimate was designed in 1954 successfully by Joe Engelberger and George Devol. It was then used in the assembly line of General Motors. It could perform various dangerous tasks with the help of an arm that it consisted of. Then, in 1956, Unimation was formed which was the first team that developed robots. These are used in automobiles and 1966, "Shakey" was developed which was the first ever mobile robot created at Stanford at the artificial intelligence center. In 1980, Dr. Eric Drexler published numerous articles based on nanoscale devices, and later in 1986, he published his book titled "Engines of Creation", the first book ever on nanotechnology [32].

2. Composites

A composite can be defined as a material that is made from two or more kinds of materials which when combined result in a material by itself when compared to the individual materials. Polymeric composites are the only composites that have tremendous applications. Electroactive polymer-based composites have been used as one of the key materials required in the making of nano-robots. All these electroactive polymer-based composites include ionic polymer-metal composites, some conductive materials are filled polymer composites, magnetic and electronic polymer gel composites, etc. that are used for the process of manufacturing the body parts of nano-robots, their sensing parts, and artificial organs which include actuator system, artificial muscles, biomimetics, etc. Some of the majorly used composites that are required for the design of the nano-robots are described below.

2.1 Liquid crystal elastomers

Liquid crystal elastomers (LECs) are those polymeric networks that have a liquid crystal moiety that causes the process of shaping morphine to occur. There is the formation of ordered domains of the liquid crystals that are anisotropic. There is a transition of domains between an isotropic state and a disordered state mainly due to the response obtained from the photo-switchable chemical unit or maybe due to the thermal transition leading to a change in shape macroscopically. There is a change in shape that can be easily manifested in the form of a linear motion which acts as an artificial muscle as well as it can be manipulated to change the shape leading to transformation in a much more complicated way. The complex solution in the shape can be attained by the process of additive manufacturing that enforces a shear that supports the domains of liquid crystal. But the complexity of the change in shape which is not at all limited to the fat sheets can be easily formed through the alignment in the domains in the presence of a magnetic field while the processing is taking place. Therefore, the skill to cause a change in the complex shapes acts as a benefit of utilizing the liquid crystal elastomers for actuation as depicted in Figure 2 [32].

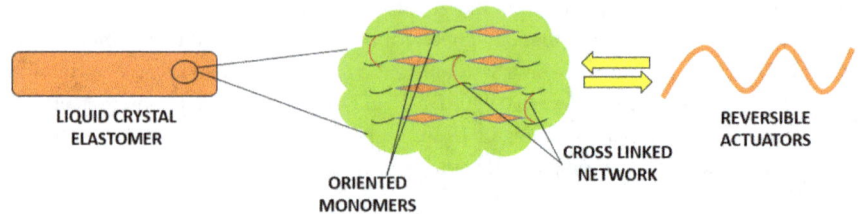

Figure 2: Schematic of the reversibility related to the liquid crystal elastomers.

2.2 Shape memory polymers

Just like liquid crystal elastomers (LCE), there exist shape memory polymers (SMPs) that can provide the capacity or capability to change the shape with a great recoverable strain that involves the usage of a compliant material. The mechanism involved in the change depends on the transition of phase which involves the crystalline domain formation or maybe a glass transition to take place. The synthesis of shape memory polymers gives a permanent change to the shape. The programmed shape is the shape through which there is cycling back of the material and moving forth towards attaining the permanent shape. In several cases, the programmed shape can be easily produced by causing deformation in the material mainly above its melting point to obtain those kinds of domains that are crystalline. These are then finally logged in a programmed shape and by the action of entropy, the permanent shape can be retained by causing melting only under stress-free conditions. One of the best qualities of shape memory polymers is that these are heat shrink tubing films that can show a one-way shape memory effect that eventually transforms to a permanent shape from a programmed shape as shown in Figure 3 [32].

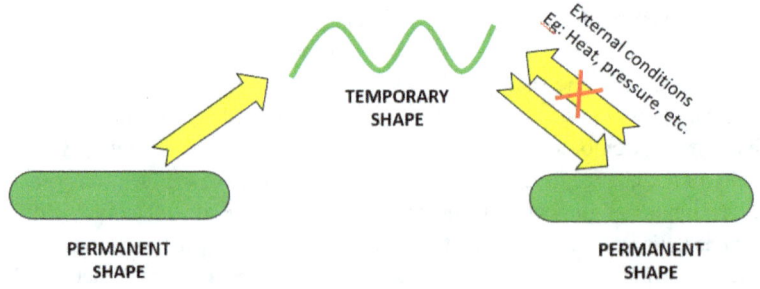

Figure 3: Schematic diagram of the nature of shape memory polymers.

2.3 Hydrogels

Hydrogels are one of the best and most useful matrix materials which have been used for functional composites. These materials have the functionality to exhibit ionic conductivity, processability, and adhesion and are very much responsive to pH, light, chemicals, ions, etc. There are certain responsivities discovered for hydrogels and one of them is the change which is caused by gelation along with the temperature change. The process of gelation can occur as the temperature decreases or increases but it also depends upon the structure of hydrogels. Hydrogels have been known to bring about a great range of functionalities into the material as shown in **Figure 4**. These can be synthesized or processed to form double networks which can improve certain properties like the toughness and stretchability of the material. These are very much essential for applications including soft robotics and even health care that include actuation, responsivity, and sensing. Hydrogels become the primary examples of those materials which have physical intelligence as soon as these are processed as composites [33].

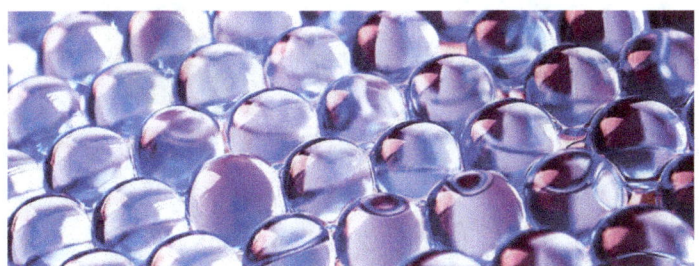

Figure 4: Orientation of a hydrogel. Adopted from the website of micropore technologies (https://microporetech.com/membrane-encapsulation/hydrogels).

2.4 CNT actuators

The single-walled carbon nanotubes (SWCNTs) can be considered as graphite having only one layer and is rolled in such a way that it forms a cylinder having a diameter of nanometers. Multiwalled carbon nanotubes (MWCNTs) have mechanical properties exceptionally good whereas SWCNTs have a tensile modulus of 640 GPa that approaches diamond, but their tensile strength is thought to be around 20-20 GPa which is around 10 times greater than any kind of the continuous fiber. The mechanical properties present in this kind of range are vividly noticed for individual single-walled carbon nanotubes but those which are noticed for the group of such nanotubes in the form of sheets are a kind of much lower that eventually restricting the actuators' performance based on the sheets of nanotubes. There is a supply of voltage between the actuating nanotube electrode along with a counter electrode with the help of a solution that contains ions which ultimately leads to charging. There exists some electrostatic force of repulsion between the charges that develop on the carbon nanotubes which toil against these stiff carbon-carbon bonds to expand and also to elongate in the form of nanotubes with the help of quantum mechanical effects which can predominate the electrostatic force is even at low levels of the injection of charges. Just like the dielectric elastomers, the strains are very much low because carbon nanotubes are extremely stiff as shown in Figure 5 [33].

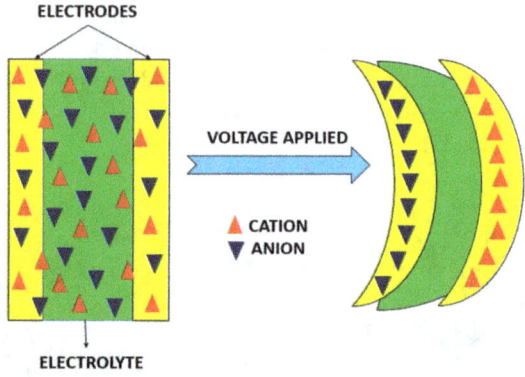

Figure 5: Schematic diagram of the actuators of carbon nanotubes.

2.5 Conducting polymers

The undoped conducting polymers are semiconductor types of materials that become conducting on doping with some donor or acceptor ions. Doping is a process that involves the addition of any impurity that improves the conductivity of a material, and it can be achieved either electrochemically or chemically. It involves various applications like electrochemical windows, energy storage, actuation devices, polymer light-emitting diodes, and sensing. Polyaniline and polypyrrole are the two best examples of common conducting polymers and their structures are shown in Figure 6. The process of insertion of an electron into an electrode could be facilitated by causing an increase in the volume with an increase in the insertion of cations as well as a decrease in the volume is observed when the anions are removed. This even depends upon the type of conducting polymer or the conducting electrolyte system that is being used, its initial state as well as the change in the potential rate that is employed for actuation. Such processes can even take place at the counter electrode and the advantage of using conducting polymers instead of electroactive polymers is that the operating voltage is low. These are cheaper than carbon nanotubes and have a high strain and low electromechanical coupling just like carbon nanotubes [34].

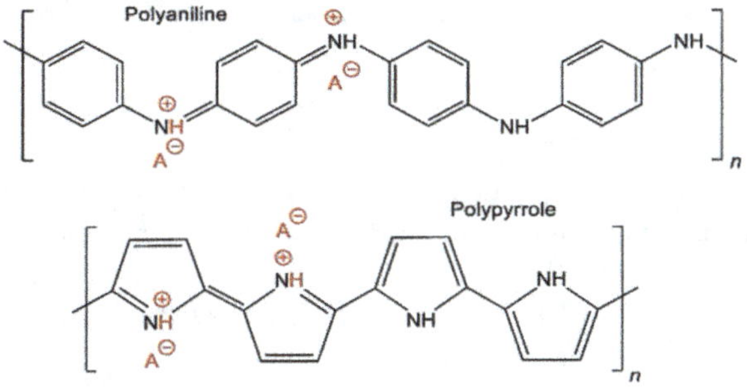

Figure 6: Chemical structures of conducting polymers: Polyaniline and polypyrrole.

3. Components and materials

Nano-robots are artificial organisms that are almost present everywhere and have a wide range of applications that have made the life of humans much easier than ever before. The nanorobots are made up of basic components and materials. Some of the main components are described below:

- **Payload:** This section comprises a very small medicine dose that can transverse in the human blood and could distribute the captured drug accurately and transfer to the particular site of injury or infected site.
- **Micro-camera:** A nanorobot has a tiny camera inside which facilitates the operator to steer the nanorobot when it navigates manually inside the body.
- **Electrodes:** The electrodes are fixed on the nanorobots and could make a battery with the help of electrolytes present in the blood. These electrodes can destroy the cancer cells through the electric current generation and hence heating the cells.
- **Ultrasonic signal generators:** These are those generators that come into play when nanorobots target and could eliminate or destroy the stones present in the kidney.
- **Swimming tail:** The nanorobots need a means through which they can enter a human body so that they could move with the bloodstream flow inside a human body.
- **Motor:** Nanorobots have motors that enable them to cause movement inside the body and these are required for the mobility of the nanorobots [34].

- **Biochip:** It is required for the manufacturing of nanorobots that can be incorporated into electronic devices that can allow teleoperations which act as the basic material mainly for some medical applications.
- **Nubots:** It is an abbreviation referring to robots made from nucleic acids. It is a synthetic robot device used as a representative to include various DNA moieties [35].

4. Movement in nanorobots

One of the most important aspects of the robots at the nanoscale is their movement and the various applications required for the trajectory which leads to their movement. About three behavioral control techniques have been recorded and taken into consideration to control the motion of the nano-robots. The first approach requires the small motion of nano-robots which is termed Brownian motion, and this helps to spot a particular target by doing a completely random search. Another approach of nanorobot required for the chemical concentration intensity is mainly for E-cadherin signals. As soon as any signal is detected, that particular nanorobot is capable to calculate the intensity of the concentration gradient and hence move toward a concentration that is higher until it extends to its target. One more approach involves the release of a chemical at the target by the nanorobot which is eventually used as a guiding signal for others to find their target. The above-mentioned are the three approaches by which the nano-robots can expand across in an arbitrary manner. Another important thing to keep in mind while carrying out the movement in the nano-robots is the avoidance of obstacles which is a vital factor to take into account in the complete control strategy of the nanorobot.

Each nanorobot that is placed inside the human body can encounter the immune system which acts as an obstacle for it along with its flow which is caused in the human body. Therefore, nano-robots must be designed in such a way that they have a strategy to avoid an escape from such an immune system that acts as an obstacle for it. So, nano-robots are equipped with sensors that are capable to detect and avoid obstacles and can identify them when it comes across them. To avoid obstacles during its trajectory, a sort of self-organized trajectory is very much required to be planned. Can be of various sizes and shapes, just for the sake of simplicity all the obstacles are considered circles for the sake of representation and the structure of the obstacles this represented as follows:

Over here, consider P center as the center point, P radium as the radium, and P velocity as the velocity of the obstacle that is moving. Consider a two-dimensional coordinate system, known as the polar coordinate system. Over here, the points that are marked on a plane are defined by a certain angle and the distance which is therefore used to spot a suitable path of trajectory required for the obstacle to be avoided during the movement of the nanorobot. Therefore, all the mentioned terms can be represented as follows:

Obstacle= <P_{center}, P_{radium}, $P_{velocity}$> (1)

A nanorobot can detect or dynamic obstacles present in its way in real-time and can get information on all the obstacles which eventually helps in the determination of the trajectory of the movement. A polar equation involving the polar coordinates of the obstacle in which, r is a variable that is the radium of the obstacle and (ri, θi) is defined as the center of the obstacle can be represented as follows and a related diagram is depicted in Figure 7 [36].

$$r^2 - 2rr_i \cos(\theta - \theta_i) + r_i^2 - r^2 = 0 \qquad (2)$$

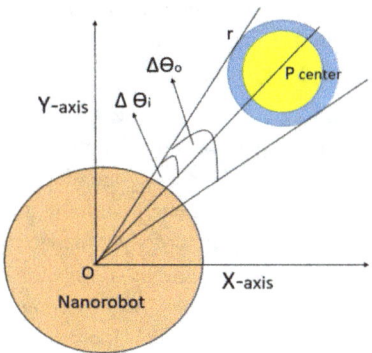

Figure 7: Schematic diagram of the obstacles in the polar coordinates.

5. Mechanism and stimulation

There are tremendous uses for nanorobots in every sphere of life. These are inserted or injected into the body of humans before acquiring a specific organ or tissue of the target. Like this, they can detect the macroscopic variations and then communicate the recorded changes to the other nanorobots present in the vicinity and thus work as a group inside the body. The target set is recognized with the help of chemical recognition because a required target has many surface chemicals that allow the nanorobots to detect and finally recognize it. Nanorobots are also able to differentiate and recognize various cells by recording the changes appearing in concentration, volume, velocity, temperature, and pressure of the cells present in the body. Nanorobot hence acts as a machine that is embedded and integrated with the devices which include data transmission, actuation, sensing, coupling power, and remote control uploading. These even have various capabilities of sensing the regions set as targets and even detecting the obstacles coming along their way [36].

The muscles are elegant machines that consist of protein filaments that interact continuously so that they could slide over one another. It is very much exciting to construct or deduce a similar kind of mechanism with the help of organic chemistry which can eventually be guided by some of the engineering considerations. These molecular actuators are being designed and are constantly investigated so that various mechanisms can be applied to both light-driven, and voltage-driven conformational changes. One of the most interesting features of the light stimulated based artificial muscles is that there is not any requirement of making contact with the actuator to carry out the energy delivery. An example of this kind of molecule is azobenzene. This has an amazing application in drug delivery which utilizes the infrared radiation of a particular wavelength that is apparent to the human body.

There is an existence of the cis- and trans-isomerization in the actuators of an excited azobenzene which causes a relative expansion in the length of the polymer which also leads to certain deformation which can be measured at the single molecular level and even macroscopically. If there is a large barrier of energy among the conformations, it reveals that the efficiency is moderate. Many alternative actuation mechanisms based on photochemical methods exist that comprise the usage of cinnamic acid-based polymers which are cross-linked reversibly when exposed to ultraviolet light. Certain challenges exist for these kinds of molecular actuators which include high quantum efficiency for actuators that are driven with photons and high electrical conductivity for those actuators which are electrochemically or electrically driven as shown in Figure 8 [37].

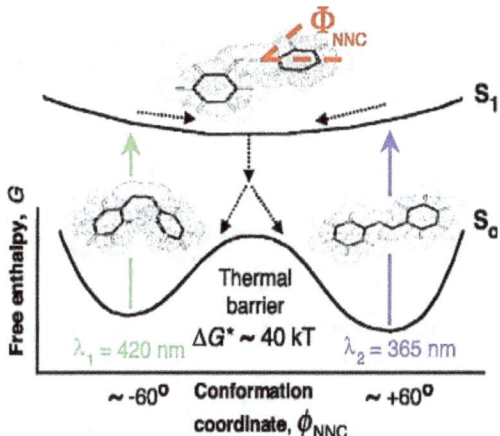

Figure 8: Diagram of cis-trans isomerism in azobenzene. Reprinted under a Creative Commons license [37].

A nanorobot operates in spotting target information in the blood vessels of the body. The fluid which is present in the vessels consists of tumor cells and red blood cells. To spot a particular target, the nanorobot utilizes the concentration of the chemical recognition of the background. If suppose any signal is not detected, then the nanorobot retains on moving along with the flow of the bloodstream in humans so that any extra power may not be utilized, and it is preserved to avoid any sort of unnecessary active locomotion. When a higher concentration is detected by a nanorobot, it can release another kind of signal which attracts the other nanorobots present in the vicinity of that nanorobot. Thus, the total nanorobots which are present in that region can be evaluated by recording the total concentration of the signal that is emitted from other nanorobots. When an appropriate coverage is recorded it then stops attracting the other nanorobots and the attracted nanorobots are then positioned as per the target area. After that, the positioned nanorobots are then moved throughout that area by detection of the higher concentration of the emitted signals. Because of the involvement of viscous forces, the nanorobots move to a new location and the movement takes place by taking alternate short routes but in a random direction. When any obstacle is rejected along the path, every nanorobot changes its position accordingly to broadcast this information to other nanorobots about the new location [37].

6. Trust dimensions

Trust is an attitude that helps in achieving the goal of an individual in a particular situation and can be characterized by vulnerability and uncertainty. It is also a reliance by any agent whose actions are well and are not at all influenced by other groups of people or organisms. For example, there is security in a bank that works upon the trust of the customers that their assets are kept safe inside it. Similarly, in artificial intelligence, it is only trust that plays a very important role in the formulation and the working process of the nanorobots. There is a total of eight trust dimensions which are described below.

6.1 Reliability and safety

The phenomenon of reliability is related to the model's capacity that could neglect the malfunctions and even the failures and can show an expected behavior or the same behavior along with time. It is even observed from both the aspects of cognitive and technical perspectives. Safety can also be examined from both technical and psychological aspects when it comes to the related dimension subject while adopting artificial intelligence and robotics. For example, accidents can be avoided through heavy-duty construction by robots.

6.2 Explainability and interpretability

In the applications of robotics and artificial intelligence, explainability is frequently related to the interpretability concept through which a system of operations can be understood by human beings with the help of explanation or introspection. It is very much important that the end users must understand the system of workflow that takes place within the input and output which is known as explainability, and along with the output, it must be understandable to the users with a proper meaning to them, known as interpretability.

6.3 Privacy and security

The most important aspect is the trust of humans in technology, and it is very much influenced by various stages of security along with privacy which is concerned with the implementation of the technology. These dimensions could be categorized under ethical and responsible artificial intelligence in which there is a need for the protection of the identity of humans along with the security of sensitive data known as the security system. These private data are protected from attackers who breach privacy and therefore, they leak private information to other communities. So, this is even regarded as a barrier to the adoption of technology where cybersecurity plays a major role. Security and privacy issues in artificial intelligence are very much known to be hampered by intelligent and automated agents which include artificial intelligence based systems like robots.

6.4 Performance and robustness

Research has depicted that there is a set of expectations that are based upon the preliminary performance of the artificial intelligence-enabled robots which gain the trust of humans. These are some of the indicators that play a vital role in developing trust trajectories. This enables to development of some expectations from humans towards nanorobots and if they fail to satisfy these kinds of expectations, then they may be responsible for losing the trust of humans. Also, robustness refers to the consistency in the performance of nanorobots in various situations in which an artificial intelligent model or the system of robots is deployed. It is very much essential in the applications as the workflow and the environmental conditions are always changing inside the body and there is a chance of failure that could occur while adapting to a new setting or the code which may be responsible for reducing or losing the trust of humans [38].

7. Actuators

Actuators are devices that can produce motion by causing the conversion of signals and energy entering into the system and this motion could be either linear or rotary. There are several types of actuators used in the field of nanotechnology that is required for making nanorobots. These are described as follows:

7.1 Thermally responsive actuators

There are some thermally responsive materials (TRM) that are first heated until they attain a particular temperature, and this is carried out by direct joule heating and can be indirectly heated with the help of ultraviolet rays. The constitution of materials at the molecular level can be altered which results in the production of the elastic modulus or deformation can occur due to contraction. Shape memory polymers have a particular elastic modulus which decreases sharply when these materials are heated above the temperature, it causes glass conversion which makes them a thermally responsive type of materials. The modulus of elasticities can be decreased by 100 times that can act as the variation in the stiffness of the robots. Shape memory polymers can be affected with the help of a temperature program, but they have an amazing advantage of adjustable stiffness and shape memory although after getting heated to the original state, they cannot go back to the program state without the help of any external force. Shape memory alloy is one of the other thermally responsive materials which can be bidirectional in the deformation state. Shape memory polymer can deform under some external forces as soon its temperature goes below the temperature at which the transition of phase takes place and is hence known as the phase transition temperature.

7.2 Photo-responsive actuators

These kinds of actuators are made by the addition of fillers that are light-sensitive to the polymers like silicone. Upon the action of light, they can expand, contract or bend. All the mentioned deformations are reversible. These kinds of actuators can be programmed to achieve a desirable motion and to have a specific response by causing a change in intensity, wavelength, and the irradiation time of the light. Also, light can be used to trigger the photochemical reactions that provide energy to the robot which helps it to drive further.

7.3 Magnetically responsive actuators

This category of actuators has been widely used as composites to utilize their ability to mix magnetic particles along with some of these materials like silicone. There can be an easy magnetization of the magnetic particles with the help of a magnetic field that ultimately generates some sort of regular magnetized curves, and their amplitude along with the direction can be easily altered. Upon the action of the magnetic field, these kinds of magnetized particles could interact with some of the spatially distributed magnetic fields and ultimately turn along with the alignment of this field. Therefore, it can be said that torque can be easily produced to bend, elongate, or shrink a complete material. It is even observed that the magnetic field could penetrate through various mediums, those actuators which are magnetically responsive are an amazing choice required for the operation caused in a very confined space. When these are compared with cable-driven robots, magnetically responsive robots are very much flexible providing a promising direction for a nanorobot to operate even in a confined or narrow space.

7.4 Electrically responsive actuators

This category of actuators can get deformed as soon as they are exposed to an external electric field. These are generally made up of two flexible electrode layers and a dielectric layer which cannot be deformed without activation and is sandwiched between the two electrode layers. When a thousand volts of voltage is applied, the electric field which is generated among the two electrodes can reduce Maxwell stress. This can compress the dielectric layer which is deformable by leading to a decrease in the thickness and hence causing an increase in the area parameter. The structure of a dielectric elastomer actuator is very much compact and various motions could be caused by a very simple actuation methodology. [39].

8. Applications

Nanorobots are artificial organism that is omnipresent and has a huge range of applications that have made the life of humans much easier than ever before. Some of the applications are mentioned below:

8.1 Cancer detection and its treatment

The nanorobots composed of a polymer mixture and transferrin as a protein can detect the tumor cells because of an amazing molecular particularity. As soon as they enter into the cells of the body, there is a chemical sensor that enables the nanoparticles to dissolve and set some free substances to actuate on the RNA of every cell which ultimately disables the genes that are causing cancer.

8.2 Nanorobots in the diagnosis and treatment of diabetes

Glucose is essential to maintain the process of metabolism in humans and it flows through the stream of blood. Its suitable level is very much essential to diagnose the level of diabetes. With an amazing immune system also inside the body, nanorobots have biocompatibility due to which they couldn't be encountered by the white blood cells. For regulating glucose levels inside the body, it utilizes a chemosensor which is embedded and determines the protein glucosensor activity.

8.3 Artificial oxygen carrier nanorobot

Respirocyte is a red cell which is also known as an artificial mechanical red cell. It is an imaginary nanorobot that glides along with the stream of blood. Outside the devices, gas concentration sensors are present. As soon as the nanorobot passes through the capillaries of the lungs, there is the development of a high partial pressure of oxygen and the partial pressure of carbon dioxide is quite low. When a high partial pressure of carbon dioxide is developed and the partial pressure of oxygen becomes lower, respirocytes then stimulate the natural hemoglobin-filled red blood cells' action [40-48].

9. Future challenges

There are certain challenges observed in the field of nanorobotics while making and using nanorobots. One of the key challenges comprising highly functional materials is that there is a huge level of length scales ranging from micro-scale to meter scale to which a material could be extended. It is this property that makes a problem in the manufacturing of nanorobots that eventually becomes a limitation [49]. As power could be supplied wirelessly, and the computational elements could be localized in the structure with the help of external fields, it ultimately poses a limitation to maintaining this wireless technology and the required backup for it. Challenges do not lie only in the development of various components but also in the interaction of nanorobotic materials. Therefore, considering the properties of the material would affect the system in which the integration could be controlled is also necessary [50].

Conclusion and future scope

Nanotechnology is an evolving tool that has a wide set of applications mainly in the field of medicinal applications. This sector mostly includes applications in dentistry, diabetes, cancer, arteriosclerosis & gene therapy. It has been clearly explained how new developments in manufacturing technologies are leading to innovative works. Also, they might help in employing and constructing the nanorobots most essentially for biomedical applications. Those kinds of materials which make the robots smart have an amazing potential to simplify robot design by causing an off-loading in the processing of signals and even leading to control of the material. It, therefore, leads to the abstraction of high-level functions. Soft machines will require multifunctional materials for versatility and robustness.

It can be very well realized that the applications of nanorobots in the field of health raise new challenges in controlling. There exist intelligent control systems designed for the nanorobots which may cause a great impact on the creation of medical nanorobotic systems required in the future. Robots or artificial organisms, embrace the qualities that make nanotechnology an amazing branch of science and arises a keen interest in the researchers to bring more advancements in the development of smart materials required for robotics that can withstand all the harsh conditions and make the world a better place to live in with the ease of science, making the lives of humans easier.

Conflict of interest

None

Acknowledgment

Authors extend their thanks and appreciation to Amity Institute of Applied Sciences, Amity University Uttar Pradesh, Noida, India for their constant support and encouragement throughout of this work.

References

[1] T.T. Kessler, C. Larios, T. Walker, V. Yerdon, P. Hancock, A comparison of trust measures in human-robot interaction scenarios, in: P.S. Knepshield, J. Chen (Eds.),

Advances in Human Factors in Robots and Unmanned Systems, Springer, 2017, pp. 353-364. https://doi.org/10.1007/978-3-319-41959-6_29

[2] Royal_Society, The Frontiers of Machine Learning: 2017 Raymond and Beverly Sackler US-UK Scientific Forum, National Academy of Sciences, 2018. ISBN: 0309471958. National Academy of Sciences. 2018. The Frontiers of Machine Learning: 2017 Raymond and Beverly Sackler U.S.-U.K. Scientific Forum. Washington, DC: The National Academies Press.

[3] N. Wang, D.V. Pynadath, S.G. Hill, Trust calibration within a human-robot team: Comparing automatically generated explanations, in Proceedings of the 11th ACM/IEEE International Conference on Human-Robot Interaction (HRI), IEEE, Christchurch, New Zealand, 2016, pp. 109-116. https://doi.org/10.1109/HRI.2016.7451741

[4] E. Glikson, A.W. Woolley, Human trust in artificial intelligence: Review of empirical research, The Academy of Management Annals. 14 (2020) 627-660. https://doi.org/10.5465/annals.2018.0057

[5] A.F. Winfield, M. Jirotka, Ethical governance is essential to building trust in robotics and artificial intelligence systems, Phil. Trans. R. Soc: A, 376 (2018) 20180085. https://doi.org/10.1098/rsta.2018.0085

[6] T. Nomura, Robots and gender, Gender and the Genome 1 (2017) 18-25. https://doi.org/10.1089/gg.2016.29002.nom

[7] L. Zhang, Y. Pan, X. Wu, M.J. Skibniewski, Artificial Intelligence in Construction Engineering and Management, Springer, Singapore, 2021. https://doi.org/10.1007/978-981-16-2842-9

[8] J.M.D. Delgado, L. Oyedele, A. Ajayi, L. Akanbi, O. Akinade, M. Bilal, H. Owolabi, Robotics and automated systems in construction: Understanding industry-specific challenges for adoption, Journal of Building Engineering 26 (2019) 100868. https://doi.org/10.1016/j.jobe.2019.100868

[9] Z. Chen, F. Gao, Y. Pan, Novel door-opening method for six-legged robots based on only force sensing Chinese Journal of Mechanical Engineering 30 (2017) 1227-1238. https://doi.org/10.1007/s10033-017-0172-7

[10] J. Carpentier, N. Mansard, Multicontact locomotion of legged robots, IEEE Transactions on Robotics 34 (2018) 1441-1460. https://doi.org/10.1109/TRO.2018.2862902

[11] V. Kaartemo, A. Helkkula, A systematic review of artificial intelligence and robots in value co-creation: Current status and future research avenues, J. Creat. Value 4 (2018) 211 228. https://doi.org/10.1177/2394964318805625

[12] D. Rus, A decade of transformation in robotics, in Towards a New Enlightenment? A Transcendent Decade, BBVA, 2018.

[13] C. Wang, K. Sim, J. Chen, H. Kim, Z. Rao, Y. Li, W. Chen, J. Song, R. Verduzco, C. Yu, Soft ultrathin electronics innervated adaptive fully soft robots, Advanced Materials 30 (2018) 1706695. https://doi.org/10.1002/adma.201706695

[14] Y. Tang, L. Qin, X. Li, C. Chew, J. Zhu, A frog-inspired swimming robot based on dielectric elastomer actuators, IEEE/RSJ International Conference on Intelligent Robots & Systems, Vancouver, Canada, 2017, pp. 2403-2408. https://doi.org/10.1109/IROS.2017.8206054

[15] S. Seok, C.D. Onal, K.J. Cho, R.J. Wood, Meshworm: A peristaltic soft robot with antagonistic Nickel Titanium coil actuators, IEEE-ASME Transactions on Mechatronics 18 (2013) 1485 1497. https://doi.org/10.1109/TMECH.2012.2204070

[16] M. Sivasankar, Brief review on nano robots in bio medical applications, Adv. Robot. Autom. 1 (2012) 1000101. https://doi.org/10.4172/2168-9695.1000101

[17] Z. Ren, W. Hu, X. Dong, M. Sitti, Multi-functional soft-bodied jellyfish-like swimming, Nature Communications 10 (2019) 1-2. https://doi.org/10.1038/s41467-018-07882-8

[18] M.P. Cunha, S. Ambergen, M.G. Debije, E.F.G.A. Homburg, J.M.J.D. Toonder, A.P.H.J. Schenning, A soft transporter robot fueled by light, Adv. Sci. 7 (2020) 1902842. https://doi.org/10.1002/advs.201902842

[19] S. Shian, K. Bertoldi, D.R. Clarke, Dielectric elastomer based "grippers" for soft robotics, Advanced Materials 27 (2015) 6814-6819. https://doi.org/10.1002/adma.201503078

[20] B. Mazzolai, L. Margheri, M. Cianchetti, P. Dario, C. Laschi, Soft-robotic arm inspired by the octopus: II. From artificial requirements to innovative technological solutions, Bioinspiration & Biomimetics 7 (2012) 025005. https://doi.org/10.1088/1748-3182/7/2/025005

[21] Q. Shen, T. Wang, J. Liang, L. Wen, Hydrodynamic performance of a biomimetic robotic swimmer actuated by ionic polymer-metal composite, Smart Materials and Structures 22 (2013) 075035 . https://doi.org/10.1088/0964-1726/22/7/075035

[22] M. Falahati, P. Ahmadvand, S. Safaee, Y.C. Chang, Z. Lyu, R. Chen, L. Li, Y. Lin, Smart polymers and nanocomposites for 3D and 4D printing, Mater. Today 40 (2020) 215-245. https://doi.org/10.1016/j.mattod.2020.06.001

[23] T. Paulino, P. Ribeiro, M. Neto, S. Cardoso, A. Schmitz, J.S. Victor, A. Bernardino, L. Jamone, Low-cost 3-axis soft tactile sensors for the human friendly robot Vizzy, in International Conference on Robotics and Automation, Singapore, 2017, pp. 966-971. https://doi.org/10.1109/ICRA.2017.7989118

[24] O.M. Wani, R. Verpaalen, H. Zeng, A. Priimagi, A.P.H.J. Schenning, An artificial nocturnal\ flower via humidity-gated photoactuation in liquid crystal networks, Advanced Materials 31 (2019) 1805985. https://doi.org/10.1002/adma.201805985

[25] P. Couvreur, C. Vauthier, Nanotechnology: Intelligent design to treat complex disease, Pharm. Res. 23 (2006) 1417-1450. https://doi.org/10.1007/s11095-006-0284-8

[26] P.B. Adamson, J.B. Conti, A.L. Smith, W.T. Abraham, M.F. Aaron, J.M. Aranda, J. Baker, R.C. Bourge, L.W. Stevenson, B. Sparks, Reducing events in patients with chronic heart failure study design: Continuous hemodynamic monitoring with an implantable defibrillator, Clin. Car diol. 30 (2007) 567-575. https://doi.org/10.1002/clc.20250

[27] T. Ohki, K. Ouriel, P.G. Silveira, B. Katzen, R. White, F. Criado, E. Dietrich, Initial results of wireless pressure sensing for endovascular aneurysm repair: The APEX trial acute pressure measurement to confirm Aneurysm Sac exclusion, J. Vasc. Surg. 45 (2007) 236-242. https://doi.org/10.1016/j.jvs.2006.09.060

[28] T. Ramcke, W. Rosner, L. Risch, Circuit configuration having at least one Nano electronics component and a method for fabricating the component 6442042US, Aug 2002.

[29] S. Das, A.J. Gates, H.A. Abdu, G.S. Rose, C.A. Picconatto, J.C. Ellenbogen, Designs for ultra-tiny, special-purpose Nano electronic circuits, IEEE Trans. Circuits Syst. - I: Regul. Pap. 54 (2002) 2528- 2540. https://doi.org/10.1109/TCSI.2007.907864

[30] R.J. Narayan, P.N. Kumta, C. Sfeir, D.H. Lee, D. Olton, D. Choi, Nanostructured ceramics in medical devices: Applications and prospects, JOM 56 (2004) 38-43. https://doi.org/10.1007/s11837-004-0289-x

[31] A. Grey, J.D. Rossiter, The Next Step - Exponential Life, 2017.

[32] A. Reghunadhan, A. Krishna, A.J. Jose, Polymers in robotics, in: M.A.A. Almadeed, D. Ponnamma, M.A. Carignano (Eds.), Polymer Science and Innovative Applications:

Materials, Techniques, and Future Developments, Elsevier, 2020, pp. 393-421. https://doi.org/10.1016/B978-0-12-816808-0.00012-3

[33] M.J. Ford, Y. Ohm, K. Chin, C. Majidi, Composites of functional polymers: Toward physical intelligence using flexible and soft materials, J. Mater. Res. 37 (2022) 2-24. https://doi.org/10.1557/s43578-021-00381-5

[34] A. Manjunath, V. Kishore, The promising future in medicine: Nanorobots, Biomed. Sci. Eng. 2 (2014) 42-47.

[35] S.S. Upadhye, B.K. Kothali, A.K. Apte, A.A. Kulkarni, V.S. Khot, A.A. Patil, R.N. Mujawar, A review on nanorobots, Am. J. PharmTech Res. 9 (2019) 11-20. https://doi.org/10.46624/ajptr.2019.v9.i2.002

[36] K. Haymar, S. Hla, Y. Choi, J.S. Park, Obstacle avoidance algorithm for collective movement in nanorobots, J. Comput. Sci. 8 (2008) 302-309.

[37] T. Mirfakhrai, J.D.W. Madden, R.H. Baughman, Polymer artificial muscles, Mater. Today 10 (2007) 30-38. https://doi.org/10.1016/S1369-7021(07)70048-2

[38] N. Emaminejad, R. Akhavian, Automation in construction trustworthy AI and robotics: Implications for the AEC industry, Autom. Constr. 139 (2022) 104298. https://doi.org/10.1016/j.autcon.2022.104298

[39] Y. Hao, S. Zhang, B. Fang, F. Sun, H. Liu, H.A. Li, Review of smart materials for the boost of soft actuators, soft sensors, and robotics applications, Chinese J. Mech. Eng. 35 (2022) 37. https://doi.org/10.1186/s10033-022-00707-2

[40] S.L. Uriarte, Low dimensional systems nanorobots, Univ. Del País Vasco, 2011.

[41] X. Yang, T. Zhou, T.J. Zwang, G. Hong, Y. Zhao, R.D. Viveros, T.M. Fu, T. Gao, C.M. Lieber, Bioinspired neuron-like electronics, Nat. Mater. 18 (2019) 510–517. https://doi.org/10.1038/s41563-019-0292-9

[42] S.P. Lacour, G. Courtine, J. Guck, Materials and technologies for soft implantable neuroprostheses, Nat. Rev. Mater. 1 (2019) 1–14. https://doi.org/10.1038/natrevmats.2016.63

[43] M. Ashrafizadeh, R. Mohammadinejad, S.K. Kailasa, Z. Ahmadi, E.G. Afshar, A. Pardakhty, Carbon dots as versatile nanoarchitectures for the treatment of neurological disorders and their theranostic applications: A review, Adv. Colloid Interface Sci. 278 (2020) 102123. https://doi.org/10.1016/j.cis.2020.102123

[44] A.V. Singh, P. Laux, A. Luch, S. Balkrishnan, S.P. Dakua, Bottom-up assembly of nanorobots: Extending synthetic biology to complex material design, Front. Nanosci. Nanotechnol. 5 (2019) 1-2. https://doi.org/10.15761/FNN.1000S2005

[45] X.F. Ye, Y.N. Hu, S.X. Guo, Y.D. Su, Driving mechanism of a new jellyfish-like microrobot, in Proceedings of IEEE International Conference on Mechatronics and Automation, China, 2008, pp. 563-568.

[46] J. Najem, D.J. Leo, A bio-inspired bell kinematics design of a jellyfish robot using ionic polymer metal composites actuators, in Proceedings of SPIE, Blackburg, USA, 2012. https://doi.org/10.1117/12.915170

[47] S.S. Nakshatharan, V. Vunder, I. Poldsalu, U. Johanson, A. punning, A. Aabloo, Modeling and control of ionic electroactive polymer actuators under varying humidity conditions, Actuators 7 (2018) 7. https://doi.org/10.3390/act7010007

[48] M. Sitti, Physical intelligence as a new paradigm, Extreme Mech. Lett. 46 (2021) 101340. https://doi.org/10.1016/j.eml.2021.101340

[49] R. Pfeifer, J. Bongard, How the Body Shapes the Way We Think: A New View of Intelligence, MIT Press, Cambridge, MA, 2007. https://doi.org/10.7551/mitpress/3585.001.0001

[50] D. Hughes, N. Correll, C. Heckman, Materials that make robots smart, The International\ Journal of Robotics Research 38 (2019) 1338-1351. https://doi.org/10.1177/0278364919856099

Chapter 2

Data Mining in Material Science

Moganapriya Chinnasamy[1], Rajasekar Rathanasamy[2*], Samir Kumar Pal[1], Manoj Kumar Kathiresan[2], Sathish Kumar Palaniappan[1]

[1]Department of Mining Engineering, Indian Institute of Technology Kharagpur, West Bengal State, India

[2]Department of Mechanical Engineering, Kongu Engineering College, Erode, Tamil Nadu State, India

* rajasekar.cr@gmail.com

Abstract

The invention of novel materials is the most propelling factor promoting the growth of contemporary civilization and technological innovation; nevertheless, previous materials research relied mostly on random techniques, which is arduous and labor-intensive. With the introduction of big data, which brings a deep upheaval in human society and considerably advances science, artificial intelligence, machine learning, and deep learning methods have recently made remarkable progress in materials science research. However, there are few comprehensive generalizations and descriptions of its applications in materials research. In this chapter, a brief summary of the evolution of materials science research is presented, followed by an emphasis on the key principles and basic processes of AI technique.

Keywords

Artificial intelligence, Machine learning, Deep learning, Materials

Contents

Data Mining in Material Science 24

1.   Introduction 25
2.   Machine learning and materials science 28
3.   ML algorithms in materials science 31

4.	Steps in machine learning for materials science	33
	4.1 Experience	33
	4.2 Task	34
	4.3 Classification	34
	4.4 Regression	34
	4.5 Clustering	34
	4.6 Dimension reduction and conception	34
	4.7 Efficient searching	35
	4.8 Performance measure	35
	4.9 Model particulars	35
	4.10 Supervised model	36

Conclusion 38

References 39

1. Introduction

Artificial intelligence (AI) is a branch of computer science that aims to understand and build intelligent systems. Researchers have seen remarkable development in AI in recent decades, with some systems already achieving human-level and potentially super-human competence for a variety of activities such as voice recognition, picture analysis, machine interpretation, and games. There is a widespread conviction that AI will fundamentally alter many aspects of our culture and economy. Self-driving vehicles, for example, which incorporate actual visual identification and control, are on the verge of becoming a reality. This incredible development is causing a major transition in AI exploration from theoretical pursuit to an abundant larger sector with significant industry and governmental funding.

Considering the remarkable breakthroughs in AI, the research fraternity as a whole has taken notice and is investigating the application of AI for discoveries [1-11]. The materials science sector, in particular, has begun to use AI tools to speed material discovery. The current tendency is to apply machine learning (ML) approaches because many recent AI successes, particularly those involving superhuman powers are anchored in ML and deep learning (DL) approaches. Even though the conditions that facilitate DL are indisputable, particularly for visualization, language recognition, linguistic conversion, and independent driving, the constraints are soundly identified. In general, DL methodologies greatly rely on the accessibility of massive quantities of references or labeled data that is frequently unavailable. The present state of deep learning was likened to System 1. It comprises human's usual data processing because it is quick, easy, and a form of automatic pattern identification. Perception, encompassing vision, and hearing, is a component of System 1. The perceptual capacities are extremely advanced and a significant portion of the human intellectual cortex is dedicated to observation as depicted in figure 1.

Humans as well have an additional logical System 2, which is sluggish and requires significant thought and reasoning to solve difficult issues beyond reflexive reactions unlike System 1. Such complicated jobs are unsuitable for pure ML and DL. Nonetheless, AI spans a wide range of approaches, including hunting, cognitive, preparation, and data illustration. These strategies played a variety of parts throughout the AI growth stages. They are positioned to become more significant and play a growing role in supplementing pure ML techniques as well as further extending and propelling AI research to address increasingly complicated issues. These challenges include those involving scientific discovery, as well as jobs that need critical thought and reasoning and are carried out by people utilizing System 2.

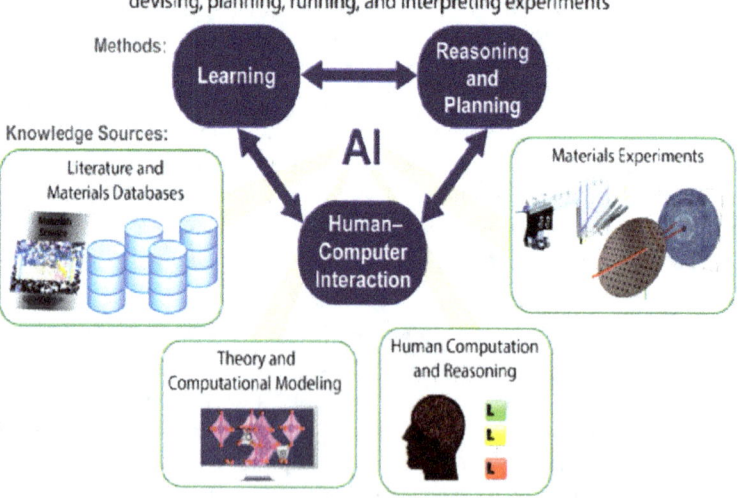

Figure 1 Structure of AI [4]

Before data-driven methodologies, material discovery evolved through three paradigms [12]. The first framework is based on experiments using trial approaches, the next is on the conversion of investigational data into a theoretical model, and the last is based on the computational simulation of theoretic representations. Individually model emerged in response to the outcomes of the preceding models. Recently, the fast expansion of existing data created through the preceding three models, along with the advent of the extremely effective algorithm, has permitted the use of AI, kicking off the quarter and the most current model of material discoveries. The main benefit of data-assisted and driven discovery is significantly reducing the necessary computing power, hence removing the limitations of the compositional transmission area. In recent years, artificial intelligence [10, 13-18] and virtual screening [19-27] approaches have been effectively applied to subfields of material and chemical science to identify novel organic compounds and constituents with extraordinary functions. Every area, due to its structural features, necessitates specialized approaches for simulated screening. Furthermore, to employ AI approaches in various sub-fields, a substantial quantity of great reliability and eminent data [28] must first be produced.

Two-dimensional materials (2D) are developing a sub-field of materials research. 2D materials, with their outstanding and controllable characteristics, show great potential for semiconductor, energy, and health applications [29-32]. Few unique 2D materials were fruitfully produced [33] later the finding of graphene [34], which has a meek 2D assembly of carbon atoms, however enticing and complicated physics. Despite a small number of experimentally produced compounds, novel 2D material depositories established on realistic considerable imitations have lately emerged [19, 35, 36]. The silicon repositories were created using two consecutive ways to build novel 2D materials: layer exfoliation through 3D bulk assemblies and combined atom interchange in 2D configurations. The initial method relies on the screening of exfoliated covered resources through 3D material databases [37-39]. This technique produces novel 2D material with assemblies that may utilize 2D patterns. The subsequent method encompasses exchanging single or additional atoms of an identified 2D pattern with molecules of extra chemical constituent. The following approach produces novel 2D constituents with identical crystal patterns as the original but with dissimilar chemical components. Regardless of the dual methodologies, the resource-intensive computation constrained the chemical search area for new 2D materials to the point where the quantity of 2D materials that were considered with density functional theory (DFT) calculations, the workhorse computational technique [40], has been reduced to a few thousand in recent years. Even though the DFT-calculated 2D materials records contain a few thousand materials, they are excellent equitable information repositories for data-driven algorithms. Amongst the primary aims, machine learning is a promising way of exploring extraordinarily broad search spaces for possible chemical combinations in the order of particles, 2D and 3D materials. The procedure creates contestant resources systematically throughout an unparalleled chemical universe of molecules, recognizes probable stable resources, and forecasts important attributes of constant materials alternatives. They utilized this technique to create a computer-generated 2D Materials databank that comprises 316,505 probable constant 2D materials with projected attributes as a proof of concept [41].

2. Machine learning and materials science

"Machine learning" implementation approaches ML approaches which contain "supervised, unsupervised and reinforcement learning" and are an essential aspect of artificial intelligence. Each instance in supervised learning is made up of input and output, often referred to as a supervised signal. The labeling of data, comprising data grouping, information attributes, and feature idea placement, is one of the major aspects of supervised learning [42]. After training the algorithms with labeled information, factors of procedures are altered and predicted on the assessment outcomes of expected ones and predicted data and the process is repeated till the procedures congregate to optimal result, at which point the intelligent decision-making capability with the specific model is obtained. Unsupervised learning, unlike supervised learning, uses unlabelled data to search and derive probable links between samples. Clustering and dimensionality reduction are two common unsupervised learning methods. Clustering requires analysis of dispersal in feature areas of data measurements to categorize analogous information into single clusters and distinct information with dissimilar types. This is necessary because the categorization of data is unclear in advance. Furthermore, because higher dimensional information cliques are fairly uncommon in ML, difficulties such as distance computations and insufficient sample data are common, which is referred to as dimension catastrophe. The theoretical basis of reducing the dimension in ML is plotting the information sockets from novel higher dimensional to lower dimensional area and dimension reduction procedures primarily include factor analysis (FA), singular value decomposition (SVD), principal component analysis (PCA), and independent component analysis (ICA) [43-47].

Unlike "supervised and unsupervised learning, reinforcement learning" obtains knowledge statistics and updates the typical factors continuously based on input from the situation, it does not necessitate information to be supplied prior. In general, when the machine performs appropriately, a positive incentive is obtained; otherwise, a negative incentive is gathered. In this situation, the ML algorithms create several dynamic planning concepts, and eventually picked the action mode which maximizes the incentive. This demonstrates that the reinforcement learning technique requires less information and is easier to develop, making it suitable for dealing with increasingly complex choice issues. Furthermore, as shown in figure 2 deep reinforcement learning that associates reinforcement through deep learning, is emerging as the fastest research hub in AI, particularly in automated driving, robots, language dispensation, and other domains [48-52]. Figure 3 depicts a framework of materials discovery which is used basically and designed based on ML approaches, with three primary processes listed: sample fabrication, algorithm model development, model verification, and materials prediction [53].

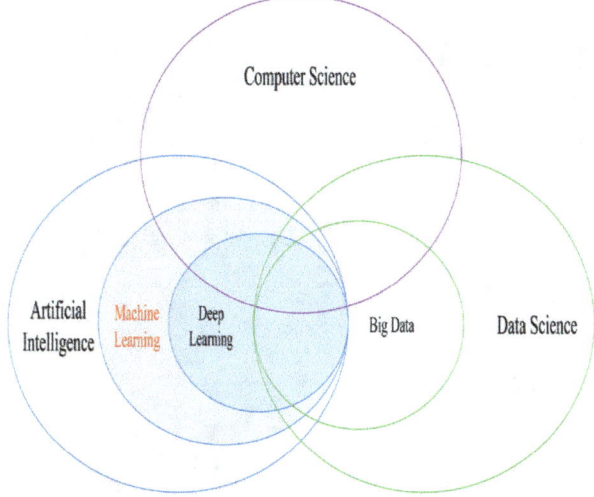

Figure 2 Interlinking of AI, ML, and DL [53]

Figure 3 Structural design of materials discovery and ML [53]

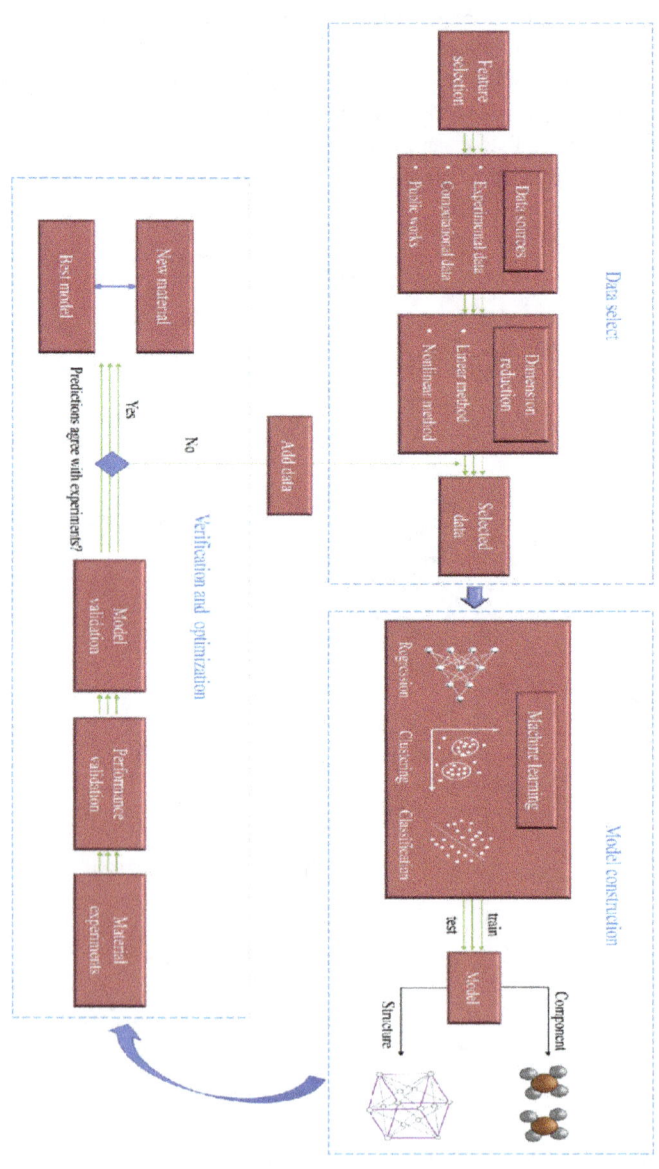

3. ML algorithms in materials science

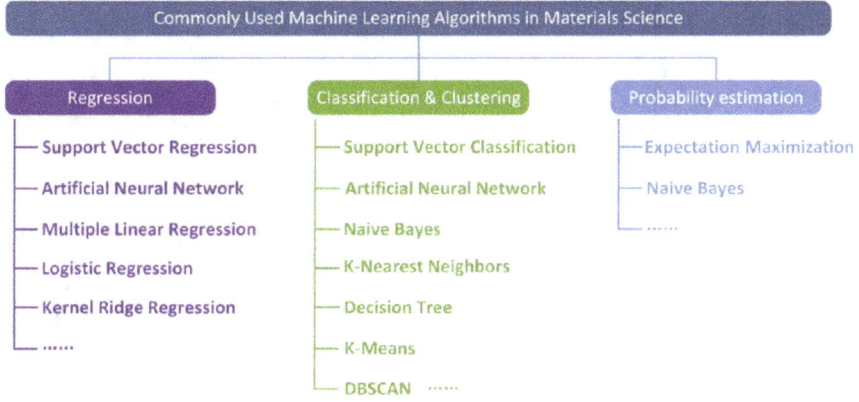

Figure 4 ML algorithm in materials science [54]

The critical stage in the development of a machine learning system is the selection of an efficient machine learning system since it has a significant impact on generalization ability and prediction accuracy [55]. Every single algorithm has its unique set of applications; hence no approach is ideal for all problems. As seen in Figure 4, in materials science, the most often employed machine learning algorithms fall into 4 groupings: "probability estimation, regression, clustering, and classification" [54]. Probability estimate procedures are mostly utilized for novel material detection whereas "regression, clustering, and classification" techniques are primarily employed for predicting the material characteristics at all levels. Furthermore, ML approaches are frequently integrated with intellectual optimization techniques [56], which are primarily employed for optimizing model characteristics. Additionally, these techniques may be used to tackle additional complex tasks [57], like spatial configuration and material property optimization [58, 59] as portrayed in Figures 5 and 6.

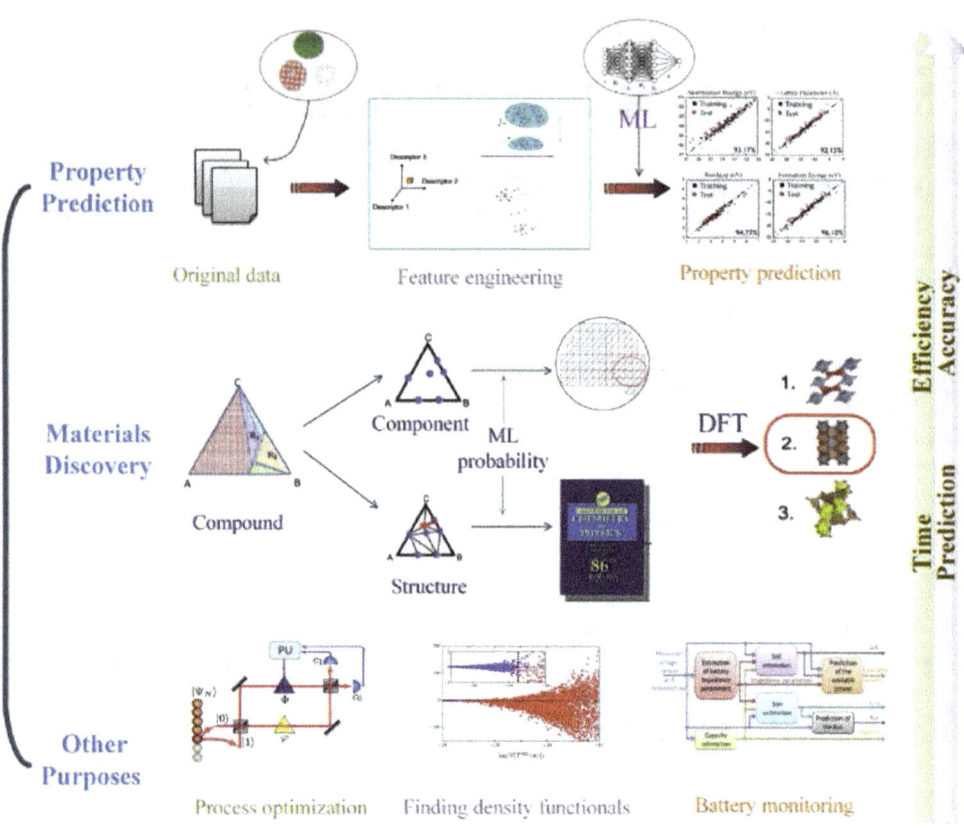

Figure 5 Application of ML in materials science [54]

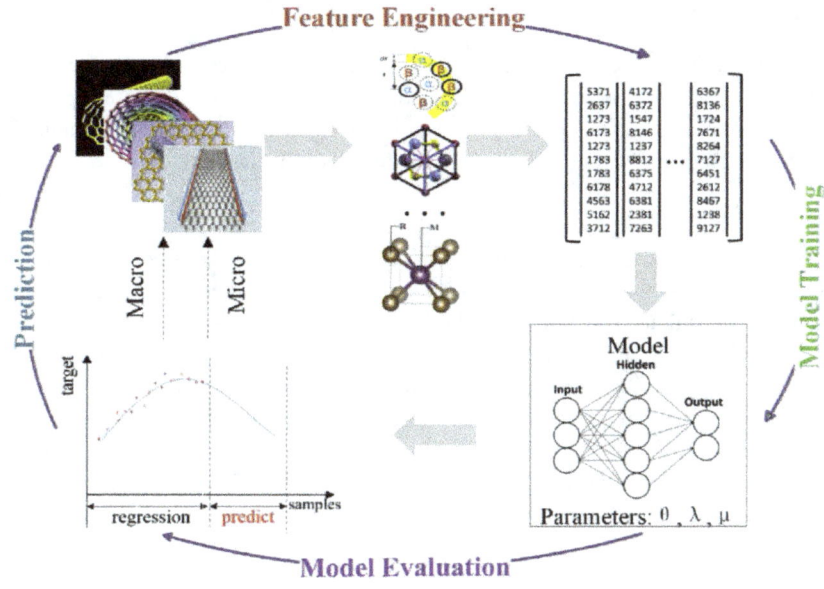

Figure 6 ML in material property forecast [54]

4. Steps in machine learning for materials science

Mitchell defines ML as a "mainframe program is supposed to acquire from experience E regard to a set of tasks T and performance measure P" [60] that is utilized as foundation background L in the following way as depicted in Figure 7.

4.1 Experience

Utmost ML algorithms may be classed as either unsupervised or supervised learning depending on the sort of samples that are provided throughout the training phase. Unsupervised learning uses a collection of instances containing just features to establish a few significant connections between the cases. A dataset of instances containing characteristics and associated labels that are "right" values linked with characteristics is used in supervised learning. Another type is reinforcement learning (RL), in which the representative knows through engaging with surroundings to acquire reward feedback. Conversely, RL confronts several obstacles, the most significant of which is the difficulty in creating interactive settings with rapid response. As a result, it is not yet commonly used in material discovery, and this evaluation confines itself to be guided and unsupervised.

4.2 Task

ML could be used for a variety of purposes. The task generally entails processing samples provided to an algorithm. The illustrations include features/descriptors that are typically organized as vectors to define the physicochemical characteristics, structural features, composition attributes, or formulation process conditions of the material.

4.3 Classification

The algorithm is concerned with determining which group or category a certain problem fits. It is accomplished through the training function f that transfers the characteristics vector to one of the unique modules. In its place of settling on a single group, f can instead provide a probability distribution across every group, with every element in the output vector representing the likelihood that the sample corresponds to the assured group. Material categorization challenges have been successfully solved using ML algorithms. A grouping model can forecast the effective formation of synthesized materials [61] and also the regions which exhibit any flow defect whenever given for specific synthesis parameters [62].

4.4 Regression

Regression is another typical job in which the algorithm attempts to study the function f= Rn/R that generates an uninterrupted y value or a series of data points represented as a vector y. Regression models are commonly used to estimate material attributes like specific heat of objects, wave equation temperature, and optical gap of materials [63-65].

4.5 Clustering

It is a type of unsupervised learning technique, which comes in handy with a huge volume of unlabelled data [66-68]. It groups the set into clusters wherein elements inside the same group are more "similar" to one another than to another. What is meant by "similar" varies depending on context and circumstances. Even if the dataset is unlabelled, grouping objects into clusters allows for valuable insight into the data. Clustering algorithms have been used in material discovery in materials textual data mining and micro-structural image processing.

4.6 Dimension reduction and conception

While teaching a network, it may depend on several variables [69]. In such cases, it is advantageous to decrease the measurement of characteristics by projecting into smaller dimensions while retaining quite as abundant data as feasible [70]. It will assist in improving computational efficiency, and model performance, reduce overfitting, and aid in the discovery of insights for specific applications. Dimension reduction was utilized to enhance prediction outcomes in material discovery, such as summarizing entire long unique features into lesser information to obtain better performance [71, 72]. Dimension reduction was employed to discover the essential science of physical prototypes by removing less important aspects [73]. Furthermore, mapping higher dimensional information to 2D or 3D charts for conception is a significant dimension reduction function since it allows meaningful insights to be drawn from an understandable graphic. It has been used to depict the high-dimensional material design space in material discovery [74].

4.7 Efficient searching

Conducting replications and lab-scale studies to collect more data is arduous and expensive. Effectual searching strategies can assist in identifying the best relevant additional data points to annotate, reducing total data-collecting efforts. Effectual probing was applied to the design of Perovskite, multilayer films, carbide, and nitride, among other materials [74, 75]. Overall, determining which objective to tackle is the initial stage in applying new ML in material discovery. Though, the five work types are not necessarily independent of one another.

4.8 Performance measure

The performance metric is employed to evaluate the performance of an algorithm on a certain job. The accuracy and error rate can be used to assess the algorithm's success in classification [76]. The log probability may be determined that produce probability dissemination. Mean square error or various variants of error are commonly used in the regression. A task's dataset is often divided into three parts: training data, a validation set of data, and data for testing. The algorithms will be taught during training and then fine-tuned based on their efficiency. The test efficiency of the optimized learning method is generally tested depending on the capability to execute test data, which functions as a measure of how effectively the model can generalize. The clustering technique in unsupervised learning may be quantified using exterior and interior indexes. Exterior indices require previously defined clustering configuration, i.e. the actual labeling for data in the dataset. The outcome of internal indices is assessed using quantities and attributes intrinsic to the database [77, 78].

4.9 Model particulars

The procedure of ML to employ is perspective and critical to select the proper model for a suitable task to obtain optimal effectiveness without inadequate or underfitting. Numerous ML models are often employed in the domain of AI-aided material discovery and are described in many papers. These approaches are introduced in supervised, unsupervised, and weakly supervised learning paradigms. Figure 1 depicts the entire architecture.

4.10 Supervised model

The supervised learning algorithm is initially qualified using labeled data containing N training instances. The ith training example, for instance, consists of a combination of dual vectors for structures and tickets. The purpose of the model is to learn the given function. The model features are critical to be determined by the problem context. Figure 8 depicts a typical supervised learning approach in material discovery. It can create datasets through lab trials, simulations, or pre-existing datasets. There are two types of characteristics that are often employed in material discovery internal, and external information about the material classification. The intrinsic content is determined by the qualities of materials and chemicals utilized, but extrinsic data is determined by the surroundings wherein the materials reside. The two sorts of characteristics can be utilized as raw features directly. However, before feeding the raw characteristics into the prediction model, certain additional modifications are usually made to them. Normalization is the first pre-processing transformation that may be used to convert the quantities of attributes to a comparable basis. Another alternative pre-processing alteration is dimension reduction, which is employed if there are too many distinct features, particularly when compared to the number of labeled training samples. In this instance, it may be preferable to lower the dimension of the attributes to avoid overfitting. [14, 79]. In addition to the above-mentioned pre-processing methods, domain precise transition will be performed using the same preceding model. The preceding framework executes a few raw characteristics and transfers the output to the last forecasting framework. The translation of diverse chemical structures, for example, into fixed-length signature vectors can aid in improving forecasting ability in molecular property prediction tasks. Following these steps, the various treated and unrefined characteristics may be reduced to generate the complete characteristic given vector. Choosing the characteristics to employ is the utmost important stage in guaranteeing a successful model presentation [80]. Regardless of the highest excellent model, it will perform poorly if the features supplied to it are unsuitable. Following feature processing, several supervised models may be employed to forecast the finalized tasks. The most often employed supervised learning algorithms in material discovery were outlined and outcomes are typically a sequence of interesting qualities expressed vectorially [81].

Figure 7 Types of ML methods [81]

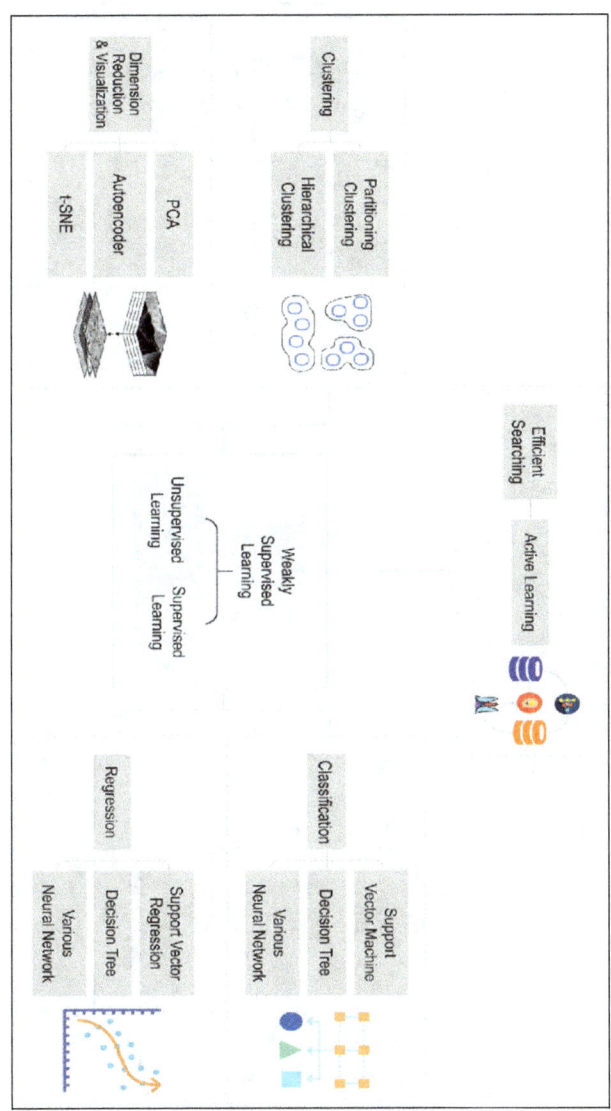

Figure 8. Supervised Learning in ML [81]

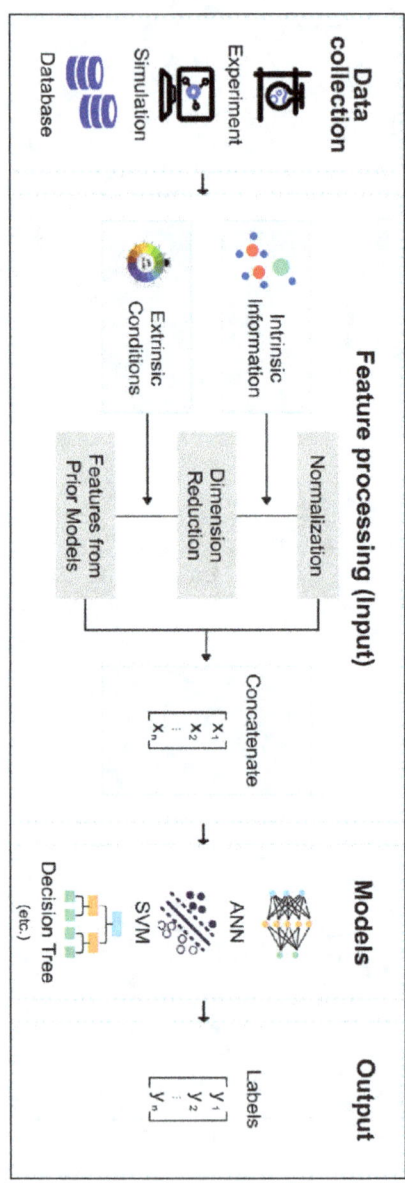

Conclusion

Machine learning, as a discipline of AI and one of the most popular categories of analytic techniques, is an important way for computers to learn. ML applications in materials science include novel material discovery, material property prediction, and other objectives spanning from the macro to micro level. The purposes of ML application in materials science are employed for quite diverse materials. A wide range of related studies shows that ML can be utilized to create precise and effective solutions for materials science. The areas in which ML can be used in materials science are becoming wider as theories and methodologies evolve.

References

[1] J. Bohannon, The cyber scientist, Science. 357 (2017) 18-21. https://doi.org/10.1126/science.357.6346.18

[2] P.D. Luna, J. Wei, Y. Bengio, A.A. Guzik, E. Sargent, Use Machine Learning to Find Energy Materials, Nature Publishing Group, 2017. https://doi.org/10.1038/d41586-017-07820-6

[3] Y. Gil, M. Greaves, J. Hendler, H. Hirsh, Amplify scientific discovery with artificial intelligence, Science. 346 (2014) 171-172. https://doi.org/10.1126/science.1259439

[4] C.P. Gomes, B. Selman, J.M. Gregoire, Artificial intelligence for materials discovery, MRS Bulletin. 44 (2019) 538-544. https://doi.org/10.1557/mrs.2019.158

[5] E.A. Hansen, S. Zilberstein, Monitoring and control of anytime algorithms: A dynamic programming approach, Artif. Intell. 126 (2001) 139-157. https://doi.org/10.1016/S0004-3702(00)00068-0

[6] A. Kelly, Think Twice: Review of Thinking, Fast, and Slow by Daniel Kahneman (2011), Numeracy. 10 (2017) 15. https://doi.org/10.5038/1936-4660.10.2.15

[7] R.D. King, K.E. Whelan, F.M. Jones, P.G. Reiser, C.H. Bryant, S.H. Muggleton, D.B. Kell, S.G. Oliver, Functional genomic hypothesis generation and experimentation by a robot scientist, Nature. 427 (2004) 247-252. https://doi.org/10.1038/nature02236

[8] J.R. Kitchin, Machine learning in catalysis, Nat. Catal. 1 (2018) 230-232. https://doi.org/10.1038/s41929-018-0056-y

[9] P. Nikolaev, D. Hooper, F. Webber, R. Rao, K. Decker, M. Krein, J. Poleski, R. Barto, B. Maruyama, Autonomy in materials research: A case study in carbon nanotube growth, NPJ Comput. Mater. 2 (2016) 1-6. https://doi.org/10.1038/npjcompumats.2016.31

[10] R. Ramprasad, R. Batra, G. Pilania, A.M. Kanakkithodi, C. Kim, Machine learning in materials informatics: Recent applications and prospects, NPJ Comput. Mater. 3 (2017) 1-13. https://doi.org/10.1038/s41524-017-0056-5

[11] E. Smalley, AI-powered drug discovery captures pharma interest, Nat. Biotechnol. 35 (2017) 604-606. https://doi.org/10.1038/nbt0717-604

[12] A. Agrawal, A. Choudhary, Perspective: Materials informatics and big data: Realization of the "fourth paradigm" of science in materials science, Apl Mater. 4 (2016) 053208. https://doi.org/10.1063/1.4946894

[13] L. Himanen, A. Geurts, A.S. Foster, P. Rinke, Data-driven materials science: Status, challenges, and perspectives, Adv. Sci. 6 (2019) 1900808. https://doi.org/10.1002/advs.201900808

[14] J. Noh, J. Kim, H.S. Stein, B.S. Lengeling, J.M. Gregoire, A.A. Guzik, Y. Jung, Inverse design of solid-state materials via a continuous representation, Matter. 1 (2019) 1370-1384. https://doi.org/10.1016/j.matt.2019.08.017

[15] Q. Zhou, P. Tang, S. Liu, J. Pan, Q. Yan, S.C. Zhang, Learning atoms for materials discovery, Proc. Natl. Acad. Sci. 115 (2018) E6411-E6417. https://doi.org/10.1073/pnas.1801181115

[16] C.C. Fischer, K.J. Tibbetts, D. Morgan, G. Ceder, Predicting crystal structure by merging data mining with quantum mechanics, Nat. Mater. 5 (2006) 641-646. https://doi.org/10.1038/nmat1691

[17] J. Schmidt, M.R. Marques, S. Botti, M.A. Marques, Recent advances and applications of machine learning in solid-state materials science, NPJ Comput. Mater. 5 (2019) 1-36. https://doi.org/10.1038/s41524-019-0221-0

[18] C. Moganapriya, R. Rajasekar, V.K. Gobinath, P. Prabhakaran, S.K. Jaganathan, A Frontier Statistical Approach Towards Online Tool Condition Monitoring and Optimization for Dry Turning Operation of SAE 1015 Steel, Archiv. Metall. Mater. 66 (2021) 901-909.

[19] J. Zhou, L. Shen, M.D. Costa, K.A. Persson, S.P. Ong, P. Huck, Y. Lu, X. Ma, Y. Chen, H. Tang, 2DMatPedia, an open computational database of two-dimensional materials from top-down and bottom-up approaches, Sci. Data. 6 (2019) 1-10. https://doi.org/10.1038/s41597-018-0005-2

[20] I.E. Castelli, K.S. Thygesen, K.W. Jacobsen, the Calculated optical absorption of different perovskite phases, J. Mater. Chem. A. 3 (2015) 12343-12349. https://doi.org/10.1039/C5TA01586C

[21] B. Meredig, A. Agrawal, S. Kirklin, J.E. Saal, J.W. Doak, A. Thompson, K. Zhang, A. Choudhary, C. Wolverton, Combinatorial screening for new materials in unconstrained composition space with machine learning, Phys. Rev. B. 89 (2014) 094104. https://doi.org/10.1103/PhysRevB.89.094104

[22] A.K. Singh, K. Mathew, H.L. Zhuang, R.G. Hennig, Computational screening of 2D materials for photocatalysis, J. Phys. Chem. Lett. 6 (2015) 1087-1098. https://doi.org/10.1021/jz502646d

[23] R.G. Bombarelli, J.A. Iparraguirre, T.D. Hirzel, D. Duvenaud, D. Maclaurin, M.A.B. Forsythe, H.S. Chae, M. Einzinger, D.G. Ha, T. Wu, Design of efficient molecular organic light-emitting diodes by a high-throughput virtual screening and experimental approach, Nat. Mater. 15 (2016) 1120-1127. https://doi.org/10.1038/nmat4717

[24] S. Curtarolo, G.L. Hart, M.B. Nardelli, N. Mingo, S. Sanvito, O. Levy, The high-throughput highway to computational materials design, Nat. Mater. 12 (2013) 191-201. https://doi.org/10.1038/nmat3568

[25] L. Yu, A. Zunger, Identification of potential photovoltaic absorbers based on first-principles spectroscopic screening of materials, Phys. Rev. Lett. 108 (2012) 068701. https://doi.org/10.1103/PhysRevLett.108.068701

[26] J. Hachmann, R.O. Amaya, A. Jinich, A.L. Appleton, M.A.B. Forsythe, L.R. Seress, C.R. Salgado, K. Trepte, S.A. Evrenk, S. Er, Lead candidates for high-performance organic photovoltaics from high-throughput quantum chemistry-the Harvard Clean Energy Project, Energy Envtl. Sci. 7 (2014) 698-704. https://doi.org/10.1039/C3EE42756K

[27] E.O.P. Knapp, C. Suh, R.G. Bombarelli, J.A. Iparraguirre, A.A. Guzik, What is high-throughput virtual screening? A perspective from organic materials discovery, Annual Rev. Mater. Res. 45 (2015) 195-216. https://doi.org/10.1146/annurev-matsci-070214-020823

[28] M.C. Sorkun, A. Khetan, S. Er, AqSolDB, a curated reference set of aqueous solubility and 2D descriptors for a diverse set of compounds, Sci. Data. 6 (2019) 1-8. https://doi.org/10.1038/s41597-019-0151-1

[29] S. Li, L. Ma, M. Zhou, Y. Li, Y. Xia, X. Fan, C. Cheng, H. Luo, New opportunities for emerging 2D materials in bioelectronics and biosensors, Curr. Opin. Biomed. Eng. 13 (2020) 32-41. https://doi.org/10.1016/j.cobme.2019.08.016

[30] N. Briggs, S. Subramanian, Z. Lin, X. Li, X. Zhang, K. Zhang, K. Xiao, D. Geohegan, R. Wallace, L.Q. Chen, A roadmap for electronic grade 2D materials, 2D Mater. 6 (2019) 022001. https://doi.org/10.1088/2053-1583/aaf836

[31] C. Moganapriya, R. Rajasekar, T. Mohanraj, V. Gobinath, P.S. Kumar, C. Poongodi, Dry machining performance studies on TiAlSiN coated inserts in turning of AISI 420 martensitic stainless steel and multi-criteria decision making using Taguchi-DEAR Approach, Silicon. 14 (2021) 4183-4196. https://doi.org/10.1007/s12633-021-01202-4

[32] C. Moganapriya, R. Rajasekar, K. Ponappa, P.S. Kumar, S.K. Pal, J.S. Kumar, Effect of coating on tool inserts and cutting fluid flow rate on the machining performance of AISI 1015 steel, Materials Testing. 60 (2018) 1202-1208. https://doi.org/10.3139/120.111271

[33] G.R. Bhimanapati, Z. Lin, V. Meunier, Y. Jung, J. Cha, S. Das, D. Xiao, Y. Son, M.S. Strano, V.R. Cooper, Recent advances in two-dimensional materials beyond graphene, ACS Nano. 9 (2015) 11509-11539. https://doi.org/10.1021/acsnano.5b05556

[34] K.S. Novoselov, A.K. Geim, S.V. Morozov, D.E. Jiang, Y. Zhang, S.V. Dubonos, I.V. Grigorieva, A.A. Firsov, Electric field effect in atomically thin carbon films. Science. 306 (2004) 666-669. https://doi.org/10.1126/science.1102896

[35] S. Haastrup, M. Strange, M. Pandey, T. Deilmann, P.S. Schmidt, N.F. Hinsche, M.N. Gjerding, D. Torelli, P.M. Larsen, A.C.R. Jensen, The computational 2D materials database: High-throughput modeling and discovery of atomically thin crystals, 2D Mater. 5 (2018) 042002. https://doi.org/10.1088/2053-1583/aacfc1

[36] M. Ashton, J. Paul, S.B. Sinnott, R.G. Hennig, Topology-scaling identification of layered solids and stable exfoliated 2D materials, Phys. Rev. Lett. 118 (2017) 106101. https://doi.org/10.1103/PhysRevLett.118.106101

[37] A. Jain, S.P. Ong, G. Hautier, W. Chen, W.D. Richards, S. Dacek, S. Cholia, D. Gunter, D. Skinner, G. Ceder, Commentary: The Materials Project: A materials genome approach to accelerating materials innovation, APL Mater. 1 (2013) 011002. https://doi.org/10.1063/1.4812323

[38] G. Bergerhoff, I. Brown, F. Allen, Crystallographic databases, International Union of Crystallography, Chester. 360 (1987) 77-95.

[39] S. Gražulis, A. Daškevič, A. Merkys, D. Chateigner, L. Lutterotti, M. Quiros, N.R. Serebryanaya, P. Moeck, R.T. Downs, A.L. Bail, Crystallography Open Database (COD): An open-access collection of crystal structures and platform for world-wide

collaboration, Nucleic Acids Res. 40 (2012) D420-D427. https://doi.org/10.1093/nar/gkr900

[40] A. Jain, Y. Shin, K.A. Persson, Computational predictions of energy materials using density functional theory, Nat. Rev. Mater. 1 (2016) 1-13. https://doi.org/10.1038/natrevmats.2015.4

[41] M.C. Sorkun, S. Astruc, J. Koelman, S. Er, An artificial intelligence-aided virtual screening recipe for two-dimensional materials discovery, NPJ Comput. Mater. 6 (2020) 1-10. https://doi.org/10.1038/s41524-020-00375-7

[42] P.R. Regonia, C.M. Pelicano, R. Tani, A. Ishizumi, H. Yanagi, K. Ikeda, Predicting the band gap of ZnO quantum dots via supervised machine learning models, Optik. 207 (2020) 164469. https://doi.org/10.1016/j.ijleo.2020.164469

[43] G.E. Hinton, R.R. Salakhutdinov, Reducing the dimensionality of data with neural networks, Science. 313 (2006) 504-507. https://doi.org/10.1126/science.1127647

[44] M. Becker, J. Lippel, A. Stuhlsatz, T. Zielke, Robust dimensionality reduction for data visualization with deep neural networks, Graphical Models. 108 (2020) 101060. https://doi.org/10.1016/j.gmod.2020.101060

[45] C. Moganapriya, R. Rajasekar, P. Sathish Kumar, T. Mohanraj, V. Gobinath, J. Saravanakumar, Achieving machining effectiveness for AISI 1015 structural steel through coated inserts and grey-fuzzy coupled Taguchi optimization approach, Structural and Multidisciplinary Optimization. 63 (2021) 1169-1186. https://doi.org/10.1007/s00158-020-02751-9

[46] C. Moganapriya, M. Vigneshwaran, G. Abbas, A. Ragavendran, V.H. Ragavendra, R. Rajasekar, Technical performance of nano-layered CNC cutting tool inserts-An extensive review, Mater. Today: Proc. 45 (2021) 663-669. https://doi.org/10.1016/j.matpr.2020.02.731

[47] G.C. Nayak, R. Rajasekar, C.K. Das, Effect of SiC coated MWCNTs on the thermal and mechanical properties of PEI/LCP blend, Composites Part A: Appl. Sci. Manuf. 41 (2010) 1662-1667. https://doi.org/10.1016/j.compositesa.2010.08.003

[48] M.N. Rastgoo, B. Nakisa, F. Maire, A. Rakotonirainy, V. Chandran, Automatic driver stress level classification using multimodal deep learning, Expert Systems with Applications. 138 (2019) 112793. https://doi.org/10.1016/j.eswa.2019.07.010

[49] L. Romeo, J. Loncarski, M. Paolanti, G. Bocchini, A. Mancini, E. Frontoni, Machine learning-based design support system for the prediction of heterogeneous machine

parameters in industry 4.0, Expert Systems with Applications. 140 (2020) 112869. https://doi.org/10.1016/j.eswa.2019.112869

[50] M. Paolanti, L. Romeo, M. Martini, A. Mancini, E. Frontoni, P. Zingaretti, Robotic retail surveying by deep learning visual and textual data, Robot. Auto. Syst. 118 (2019) 179-188. https://doi.org/10.1016/j.robot.2019.01.021

[51] P. Hähnel, J. Mareček, J. Monteil, F.O. Donncha, Using deep learning to extend the range of air pollution monitoring and forecasting, J. Comput. Phys. 408 (2020) 109278. https://doi.org/10.1016/j.jcp.2020.109278

[52] R.R. Blázquez, M.M. Organero, Using multivariate outliers from smartphone sensor data to detect physical barriers while walking in urban areas, Technol. 8 (2020) 58. https://doi.org/10.3390/technologies8040058

[53] Y. Juan, Y. Dai, Y. Yang, J. Zhang, Accelerating materials discovery using machine learning, J. Mater. Sci. Technol. 79 (2021) 178-190. https://doi.org/10.1016/j.jmst.2020.12.010

[54] Y. Liu, T. Zhao, W. Ju, S. Shi, Materials discovery and design using machine learning, J. Materiomics 3 (2017) 159-177. https://doi.org/10.1016/j.jmat.2017.08.002

[55] C.M. Bishop, N.M. Nasrabadi, Pattern recognition and machine learning, Springer, 2006.

[56] S. Fang, M. Wang, W. Qi, F. Zheng, Hybrid genetic algorithms and support vector regression in forecasting atmospheric corrosion of metallic materials, Comput. Mater. Sci. 44 (2008) 647-655. https://doi.org/10.1016/j.commatsci.2008.05.010

[57] W. Paszkowicz, K.D.M. Harris, R.L. Johnston, Genetic algorithms: A universal tool for solving computational tasks in Materials Science Preface, Comput. Mater. Sci. 45 (2009) IX-X. https://doi.org/10.1016/j.commatsci.2008.07.008

[58] C.E. Mohn, W. Kob, A genetic algorithm for the atomistic design and global optimisation of substitutionally disordered materials, Comput. Mater. Sci. 45 (2009) 111-117. https://doi.org/10.1016/j.commatsci.2008.03.046

[59] X.J. Zhang, K.Z. Chen, X.A. Feng, Material selection using an improved genetic algorithm for material design of components made of a multiphase material, Mater. Design. 29 (2008) 972-981. https://doi.org/10.1016/j.matdes.2007.03.026

[60] T.M. Mitchell, Machine Learning, McGraw-hill, New York, 1997.

[61] C.W. Coley, R. Barzilay, T.S. Jaakkola, W.H. Green, K.F. Jensen, Prediction of organic reaction outcomes using machine learning, ACS Central Sci. 3 (2017) 434-443. https://doi.org/10.1021/acscentsci.7b00064

[62] E.D. Cubuk, S.S. Schoenholz, J.M. Rieser, B.D. Malone, J. Rottler, D.J. Durian, E. Kaxiras, A.J. Liu, Identifying structural flow defects in disordered solids using machine-learning methods, Phys. Rev. Lett. 114 (2015) 108001. https://doi.org/10.1103/PhysRevLett.114.108001

[63] Y. Dong, C. Wu, C. Zhang, Y. Liu, J. Cheng, J. Lin, Bandgap prediction by deep learning in configurationally hybridized graphene and boron nitride, NPJ Comput. Mater. 5 (2019) 1-8. https://doi.org/10.1038/s41524-019-0165-4

[64] Y. Zhuo, A.M. Tehrani, A.O. Oliynyk, A.C. Duke, J. Brgoch, Identifying an efficient, thermally robust inorganic phosphor host via machine learning, Nat. Commun. 9 (2018) 1-10. https://doi.org/10.1038/s41467-017-02088-w

[65] S.K. Kauwe, J. Graser, A. Vazquez, T.D. Sparks, Machine learning prediction of heat capacity for solid inorganics, Integrating Materials and Manufacturing Innovation. 7 (2018) 43-51. https://doi.org/10.1007/s40192-018-0108-9

[66] L. Kaufman, P.J. Rousseeuw, Finding groups in data: An introduction to cluster analysis, John Wiley & Sons, 2009.

[67] M.E. Celebi, Partitional clustering algorithms, Springer, 2014. https://doi.org/10.1007/978-3-319-09259-1

[68] N. Grira, M. Crucianu, N. Boujemaa, Unsupervised and semi-supervised clustering: A brief survey, A review of machine learning techniques for processing multimedia content. 1 (2004) 9-16.

[69] A.R. Kitahara, E.A. Holm, Microstructure cluster analysis with transfer learning and unsupervised learning, Integrating Materials and Manufacturing Innovation. 7 (2018) 148-156. https://doi.org/10.1007/s40192-018-0116-9

[70] L.L.C. Kasun, Y. Yang, G.B. Huang, Z. Zhang, Dimension reduction with extreme learning machine, IEEE Trans. Image Process. 25 (2016) 3906-3918. https://doi.org/10.1109/TIP.2016.2570569

[71] T. Xie, A.F. Lanord, Y. Wang, Y.S. Horn, J.C. Grossman, Graph dynamical networks for unsupervised learning of atomic scale dynamics in materials, Nat. Commun. 10 (2019) 1-9. https://doi.org/10.1038/s41467-018-07882-8

[72] A. Mardt, L. Pasquali, H. Wu, F. Noé, VAMPnets for deep learning of molecular kinetics, Nat. Commun. 9 (2018) 1-11. https://doi.org/10.1038/s41467-017-02088-w

[73] E.Y. Lee, B.M. Fulan, G.C. Wong, A.L. Ferguson, Mapping membrane activity in undiscovered peptide sequence space using machine learning, Proc. Natl. Acad. Sci. 113 (2016) 13588-13593. https://doi.org/10.1073/pnas.1609893113

[74] K. Tran, Z.W. Ulissi, Active learning across intermetallics to guide discovery of electrocatalysts for CO2 reduction and H2 evolution, Nat. Catal. 1 (2018) 696-703. https://doi.org/10.1038/s41929-018-0142-1

[75] A. Talapatra, S. Boluki, T. Duong, X. Qian, E. Dougherty, R. Arróyave, Autonomous efficient experiment design for materials discovery with Bayesian model averaging, Phys. Rev. Mater. 2 (2018) 113803. https://doi.org/10.1103/PhysRevMaterials.2.113803

[76] B.W. Matthews, Comparison of the predicted and observed secondary structure of T4 phage lysozyme, Biochimica et Biophysica Acta (BBA). 405 (1975) 442-451. https://doi.org/10.1016/0005-2795(75)90109-9

[77] A. Thalamuthu, I. Mukhopadhyay, X. Zheng, G.C. Tseng, Evaluation and comparison of gene clustering methods in microarray analysis, Bioinformatics. 22 (2006) 2405-2412. https://doi.org/10.1093/bioinformatics/btl406

[78] S. Dudoit, J. Fridlyand, A prediction-based resampling method for estimating the number of clusters in a dataset, Genome Biology. 3 (2002) 1-21. https://doi.org/10.1186/gb-2002-3-7-research0036

[79] P.M. Shenai, Z. Xu, Y. Zhao, Applications of Principal Component Analysis (PCA) in Materials Science, in: P. Sanguansat (Eds.), Principal Component Analysis, 2012, pp. 25-40. https://doi.org/10.5772/37523

[80] M. Ayyar, M.P. Mani, S.K. Jaganathan, R. Rathanasamy, Preparation, characterization and blood compatibility assessment of a novel electrospun nanocomposite comprising polyurethane and ayurvedic-indhulekha oil for tissue engineering applications, Biomed. Tech. 63 (2018) 245-253. https://doi.org/10.1515/bmt-2017-0022

[81] J. Li, K. Lim, H. Yang, Z. Ren, S. Raghavan, P.Y. Chen, T. Buonassisi, X. Wang, AI applications through the whole life cycle of material discovery, Matter. 3 (2020) 393-432. https://doi.org/10.1016/j.matt.2020.06.011

Chapter 3

Artificial Intelligence Applications in Solar Photovoltaic Renewable Energy Systems

Ifeanyi Michael Smarte Anekwe[1]*, Emmanuel Kweinor Tetteh[2], Edward Kwaku Armah[3]

[1]School of Chemical and Metallurgical Engineering, University of the Witwatersrand, Johannesburg 2050, South Africa

[2]Green Engineering Research Group, Department of Chemical Engineering, Faculty of Engineering and the Built Environment, Durban University of Technology, Durban 4001, South Africa

[3]School of Chemical and Biochemical Sciences, Department of Applied Chemistry, C. K. Tedam University of Technology and Applied Sciences, P. O. Box 24, Navrongo, Upper East Region, Ghana

Abstract

With the increasing reduction of fossil fuel supplies, it is predicted that the world would run out of energy resources within the next few decades attributed to the depletion of fossil fuels. Renewable energy sources generate electricity with minimal contribution of CO_2 or other greenhouse gases to the atmosphere. Solar, wind, hydroelectricity, and biomass are the most common kinds of sustainable energy sources, and all of them have enormous prospects to help the world to meet its future energy demand. The solar photovoltaic technique is among the first of several renewable energy systems that have been implemented around the world to meet the basic requirement of electricity, especially in remote regions. The deployment of Artificial Intelligence in the energy sector is becoming more prevalent to ensure an effective energy supply. This chapter presents a review of the application of artificial intelligence in a solar PV system while highlighting the challenges and prospects for effective utilization in the renewable energy system.

Keywords

Artificial Intelligence, Deep Learning, Internet of Things, Machine Learning, Renewable Energy, Solar Energy, Solar Photovoltaic

Contents

Artificial Intelligence Applications in Solar Photovoltaic Renewable Energy Systems 47

1. Introduction 49
   - 1.1 Overview of Solar PV Renewable Energy System and Artificial Intelligence (AI) Technology 50
   - 1.2 Solar energy generation 53
1.3 Classification of solar energy technologies (SET) 53
   - 1.3.1 Concentrated solar-thermal power (CSP) 53
   - 1.3.2 Solar photovoltaic energy 54
2. Artificial intelligence (AI) 55
   - 2.1 Machine learning 58
   - 2.2 Deep learning 60
   - 2.2.1 Convolutional neural networks (CNNs) 61
   - 2.2.2 Long short-term memory (LSTM) 62
   - 2.2.3 Generative adversarial network (GAN) 63
3. Application of AI in solar PV system 64
   - 3.1 Monitoring of PV systems 64
   - 3.2 PV fault detection and diagnosis (FDD) methods 65
   - 3.3 Employment of AI technologies for sizing PV systems 66
   - 3.4 Modelling of a solar PV generator 66
   - 3.5 Solar water heating systems (SWHs) 67
4. Challenges of effective AI application in solar PV system 69
   - 4.1 Solar energy optimization 69
   - 4.2 PV-dependent hybrid facility optimization 71
   - 4.3 External factors of solar energy generation 72
   - 4.4 Challenges in the development of solar energy systems 73
   - 4.5 Solar energy transformation 74
5. Prospects and future work consideration 75

Conclusion 76

References 77

1. Introduction

The non-renewable global energy use has significantly raised the demand for fossil fuel resources by different sectors of the economy, resulting in a constant CO_2 emission from these sectors (Figure 1) with the energy sector contributing substantially to this emission. The increase in the impacts of global warming, climate change challenges, and a rise in world temperatures, can be linked to CO_2 emissions, which pose a prolific threat to the sustainability of the ecosystem. Because of the contaminant emissions created by non-renewable energy resources, it is projected that the world average temperature would rise by approximately 2°C by 2050 [1, 2]. To tackle these issues, timely and productive efforts must be taken to reduce adverse ecological effects while also exploring efficient and cost-effective renewable energy sources (RESs). RES techniques have been the interest of numerous studies over the last few decades, with the goal of enhancing process performance. According to a study carried out by the International Renewable Energy Agency (IRENA) in 2018 [3], the cost of electricity generated by renewable energy sources has decreased gradually over the last few years. Many nations throughout the world have benefited from this cost reduction by incorporating renewable energy sources into their national power systems. The application of renewable energy sources (RESs) to generate electricity has become increasingly popular as the world's energy consumption continues to rise as a result of the expansion of global industry and urbanisation [4].

Overall, the annual use of energy increases predictably. If the rate of global population increase continues at its current pace, the yearly utilization of oil and natural gas employed for electricity generation will increase to quadruple by 2050, as projected by the International Energy Agency (Figure 2) [5-7]. As well as these benefits, there are a variety of other reasons to shift away from petroleum toward renewable energy sources (RES), including a reduction in the costs of energy synthesis from renewable sources, a decrease in carbon emissions, a competitive market, and the consistency of RES. The application of solar and wind energies has resulted in a rise in the production of sustainable energy, with about 77 % of additional output added in 2017 [8]. Based on the International Energy Agency report [9], solar PV energy cost dropped by three-quarters between 2010 and 2017. Prices of wind turbines have decreased by around half in a comparable period, resulting in lower-cost wind energy [4].

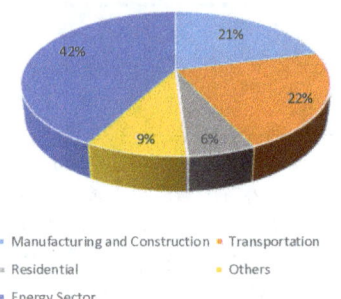

Figure 1. Global CO_2 Emission by Economic Sectors (Source IEA, [10])

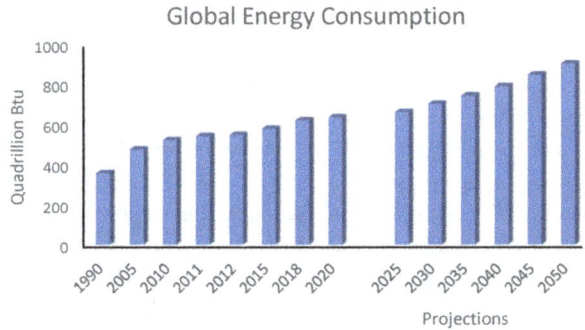

Figure 2. Global Energy Consumption from 1990 – 2020 and Predictions from 2025-2050 Source: EIA [11]

1.1 Overview of Solar PV Renewable Energy System and Artificial Intelligence (AI) Technology

The rise of solar PV energy worldwide can be broken into two parts; (i) solar PV projects and (ii) the advancements in solar PV research and development (R&D) technology [12]. On the scientific front, the number of studies discussing optimization approaches in solar PV systems has risen considerably in recent years. Also, with the publication of new academic papers, there has been an increase in the number of PV installations. According to a worldwide electricity market in 2018, solar PV, for example, surpassed all other renewable energy technologies to become the world's fastest-growing renewable energy technique [13]. The increased attention to the application of optimization technologies for solar PV system deployment is spreading worldwide, thanks to research journals published in both developed nations including the United States and Europe, and developing nations including China and India, which demonstrate how the technologies can be applied [4].

The manual techniques make use of offline operations, in which the PV module or PV string is isolated from the facility in order to diagnose and correct problem conditions. The semi-automatic and automatic techniques are executed immediately over the internet. Automatic techniques can also be divided into two major categories. The first technique seeks to build a basic algorithm to discern, recognize, and maybe localise problems based on mathematical studies, to reduce the number of steps in the process. Using artificial intelligence (AI) approaches in a more sophisticated way, the second strategy can be used to actively respond to errors, self-heal, and schedule upkeep. However, to diagnose, categorise, and pinpoint problems, a big dataset (recorded currents, voltages, pictures, solar irradiance, infrared or electro-luminance images) is used in conjunction with advanced algorithms. AI approaches have been used in solar PV farms for over two decades to enhance the modelling, control optimization, and output power prediction efficiency of large datasets [14, 15], and they continue to be used today. Other AI techniques including machine learning (ML) and deep learning (DL), are applied to manage big datasets and speed up management and decision-making [16-22]. The internet of things (IoT) facilitates interaction and data distribution among a diverse range of facilities and services. It is becoming increasingly important. In recent years, Internet of Things (IoT) techniques have also been explored in the field of PV facility surveillance and remote sensing to meet industry demand for improved fault diagnosis and prognostics [21, 22]. These researches have demonstrated that deploying IoT in the field has several advantages, including increased effectiveness and precision, reduced human involvement, and, consequently, a decrease in expenses, among others.

Commercial technology has not yet been developed that is capable of diagnosing and localising all defect types. Furthermore, they are unable to provide a precise diagnosis to trigger independent decisions promptly. For instance, determining whether a defect is serious or not, and if it necessitates an immediate response, specifically in the context of many failures, defect identification, type, and localisation in large-scale photovoltaic arrays are all important considerations. A limited number of models have been developed and studied in research laboratories. To determine whether this prototype is cheap on a large scale, knowledge dissemination from laboratories to the industry must take place before it can be implemented on a wider scale [23]. Furthermore, most of the artificial intelligence systems that have been developed have been largely confirmed in laboratory simulation, with only a handful having been truly proven empirically. This can be attributed to the difficulty in obtaining precise results that accurately simulate real-life situations, limited facilities, and the expense of deployment.

Environmental issues faced by developing nations as a result of greenhouse gases caused by an overreliance on fossil fuels are generating worldwide concerns, highlighting the need for renewable energy (RE). There is a policy in Europe, America, and emerging nations, most notably Brazil and India, that encourages the use of RE including solar photovoltaics (PV), wind energy, and biomass since these are considered clean and ecologically benign [24, 25]. The resources for RE are widely available, and numerous methods may be used in energy production. Despite the abundance of RE resources, according to the global RE assessment, RE accounted for 29 % of global power output as of 2020, up from 27 % in 2019. Renewables such as wind power, solar energy, and biomass contribute 3 % of electricity and 2 % of primary energy, accounting for a smaller portion of total energy demand. A significant amount of RE is reported to be derived from hydroelectric resources and agricultural waste, which is abundantly used in underdeveloped nations [26].

Regardless of capacity, the existing contribution of renewable resources to the global energy supply is worthwhile. Renewable resources provide approximately 17 % of the world's primary energy on average; however, 'new' renewable resources such as solar, provide a smaller quantity of energy needs at the moment – approximately 2 % of primary energy [26]. A noticeable characteristic of RE was attributed to the variety of technologies and available resources accessible with few questions such that the greatest size of the resources is huge, and that it could provide a considerable contribution to global needs, easily surpassing present global power supplies. An illustration of this assertion is presented in Table 1. RE provides several solutions to the environmental and social problems caused by traditional production methods.

Table 1. Renewable energy resources and global power supplies [27].

	Capacity (MW)	Approximate annual output (TWh/year)
Biomass	35,000	185
Wind	20,000	50
Solar thermal	350	0.2
Solar PV	1200	1
Geothermal	8,200	44
Total	68,550	
	Capacity (GW)	Output (TWh/year)
Current world electricity	3,000,000	15,000

1.2 Solar energy generation

Although the global application of RE has surfaced, the non-predictable nature of these resources has placed a significant burden on large-scale energy production projects. Müller-Steinhagen [28] assessed the problems confronting the RE industry for decades and highlighted the significance of optimizing RE laws and processes to maximize the advantages of existing power systems. Furthermore, by appropriately describing operational characteristics in RE, accurate performance optimization issues may be handled. Powell et al., [29] constructed a viable model for the circulation of power in some parts of the United States of America, specifically Carolina to develop a self-sustaining separation capacity. Previous studies have proven that if the suggested model fed accurately, it might generate more accurate conclusions with increased system performance. Indeed, in recent years, research has focused on developing an innovative and effective optimization strategy to handle RE challenges, notably for wind and solar energy.

1.3 Classification of solar energy technologies (SET)

SET to a larger extent, provides sustainability to anthropological actions [25]. A huge reduction in greenhouse gas emissions is one of the most significant environmental benefits of the SET. In this section, two major classifications are discussed: concentrating solar-thermal power and solar PV energy.

1.3.1 Concentrated solar-thermal power (CSP)

CSP is a form of heat energy that may carry electricity using sun radiation [30]. Solar radiation, also known as electromagnetic radiation, is the light released by the sun: as a result, CSP systems only employ direct sunlight; the strength of electromagnetic radiation is commonly known as the Direct Normal Irradiation (DNI). Mirrors provide a reflecting and focusing solar radiation onto receivers, which tend to receive energy and convert it to heat, which may then be utilized to create electricity or stored for future use. CSP is carbon-free, less costly, and predominantly employed in huge power plants, which increases its likelihood of being the world's future RESs technology for generating energy. CSP is entirely dependent on the extent of straight-forward electromagnetic radiation from the sun. According to studies, the CSP technique is a form of RE that, in the future, might become a globally principal electricity source [24]. Such techniques seek to offer a dependable, safer and benign power. CSP is considered favourable among other RE sources since it is carbon-free and can be generated via heat energy for storage. Because thermal energy storage may be installed in a power plant, electricity can be delivered to customers without interruption while the components of the power plant are being maintained [25]. CSP is said to lower carbon dioxide levels in the atmosphere, making it a more essential technology in the near future than fossil fuels. In the USA, notably in the Southwest, the land management bureau has established CSP projects worth suggestively more than the Western Governors Association's 2006 Solar Task Force Report, which is scheduled to last until 2015 [31]. Even though CSP plants employ a variety of CSP technologies, all of them use the same operation techniques/processes in order to generate energy.

There are four categories of CSP technologies as these systems are either commercialized or on their way to being so [32]. Parabolic troughs, solar towers, parabolic dishes, and linear Fresnel reflectors are examples of CSP systems. Based on the method of operation, these systems are further classified as linear focusing systems (focus solar radiations into parallel tube receivers arranged above a row of mirrors (optical concentrators) or point focusing systems (focus solar radiations reflected from optical concentrators around a central tower/point which functions as the receiver). Parabolic Troughs are a form of linear focusing system on which optical concentrators are designed in the shape of parabolic troughs making them capable of collecting parallel radiations as well as a single line focus. They gather and focus solar radiation onto parallel receivers situated at the parabola's focal line. Consequently, CSP technology has a minor environmental effect; compared to fossil fuels which supersede that of networks of power supplies. Temperature disparities occur from changes in the form of the technique [29].

1.3.2 Solar photovoltaic energy

The global growth of solar PV energy can be broken into two parts: The increase of solar PV projects and research, and the breakthroughs in solar PV R&D technology [12]. On the scientific front, the number of research articles published during the last decade that discuss the use of optimization technologies in solar PV facilities has expanded dramatically. In addition to the advancement of scientific articles, the number of PV installations has increased. According to a worldwide electricity market in 2018, solar PV, for example, surpassed all other renewable energy technologies to become the world's fastest-growing renewable energy technique [13]. The increased attention to the application of optimization technologies for solar PV system deployment is spreading worldwide, thanks to research journals published in both developed nations including the United States and Europe, and developing nations including China and India, which demonstrate how the technologies can be applied [4].

In the 1960s, PV was utilized to power satellites in the United States space program. Niche applications that demand an extremely less quantity of power for use in calculators and telecommunication networks were also early drivers of industry growth [33]. However, in recent years, favourable regulations from certain nations have contributed significantly to establishing PV markets for both grid and off-grid power delivery. Since the mid-1980s, PV has reached an average of 15 % per year in the past few years which rose to over 30 % and far-reaching 40 %. PV has grown from a specialized laboratory to a specialized factory, and this expansion has been related to the creation of progressively bigger dedicated manufacturing units. As a result, significant automation and standardization of industrial processes, as well as economies of scale, have occurred [34]. A serious problem that could arise from PV systems is the degree to which non-amorphous silica allows for "thin-film" strategies in energy generation. A PV power generating system consists of several compositions which include the mountings and the cells, collectively. Peak kilowatts (kWp) measure the electrical power that a system could provide directly from the sun at an overhead compartment [35]. A grid-connected system ranges from a little kWp for residential areas to solar-powered installations to hundreds of GWp. Sinhal et al., [34] utilized experience curve methods to anticipate the diverse degree of PV systems necessary to achieve various price-to-cumulative-shipments patterns. The universal solar radiation and sunshine time figures have been studied by Pundir et al., [36]. In that study, the RE generation and economic assessment for a 5 MW PV-based grid power system in Saudi Arabia were found feasible to meet global standards.

2. Artificial intelligence (AI)

Artificial Intelligence (AI) refers to data processing systems and technical resources linked to human intelligence to execute a task [37]. AI may be incorporated into hardware systems or dependent on software mostly in the virtual world [38]. However, to display human intelligent behaviour and specified goals, AI-based systems can analyse their environment and take actions autonomously [39]. Moreso, AI can also use, train, and analyse data sets for decision-makers in database analysis, accounting, information retrieval, product design, medicine, food quality monitoring, biometrics, forensics, production planning and distribution [38-40].

In environmental and renewable energy applications, AI is gaining momentum because of its capacity to automate systems for enhanced reliability and profitability [37, 39, 41, 42]. Since renewable energy data is inherently complex, AI in power systems can help to improve network stability, reliability, and dynamic performance [38, 42, 43]. Among these are optimization, data exploration, categorization, regression, and grouping [40, 42]. AI development can also promote system learning in system design, control, and maintenance to improve performance and quality [38]. This method fostered data-driven research to investigate complex and hard-striking challenges in power systems. To solve complex and ill-defined problems, some computational intelligence technologies are overtaking conventional data processing techniques [44]. Among the most effective AI techniques are knowledge-based systems including artificial neural networks (ANN) and fuzzy logic. Figure 3 shows that DL is a subdivision of ML, and thus both are AI techniques. Theories like statistics, neural networks, and evolutionary learning are all used in these AI techniques [43, 44]. The suitable approach for a given application depends on the nature of the issue, the accessibility of data, and the necessary accuracy and easiness [45, 46]. Table 2 presents AI solutions and methodologies for grid-connected PV systems.

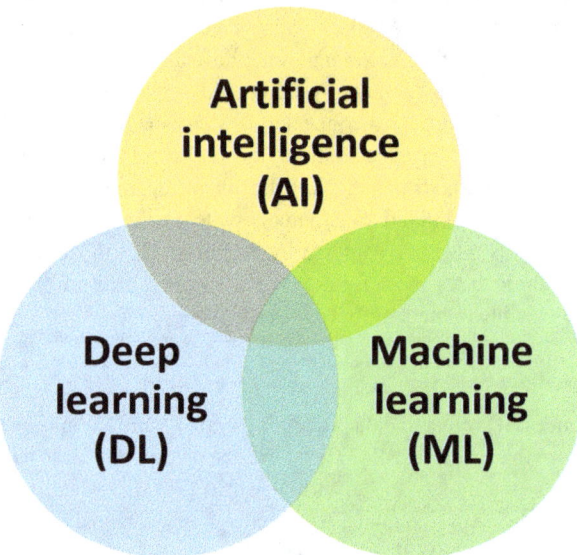

Figure 3. Venn graphic demonstrating the correlation between AI, ML, and DL adapted from Mellit and Kalogirou [38]

Table 2. Demerits of conventional algorithm application and artificial intelligence (AI) remediation adapted from Kurukuru, et al. [44]

Conventional algorithms (CA)	Application	Setbacks of CAs	Solutions with AI	AI techniques
Projecting and random techniques	Supervising and upkeep	Searching to outliers	Using quantile systems, replace outliers with reasonable values.	-Machine learning (ML) -Deep learning (DL)
Information minimisation methods	Upkeeping	Needs collection and smart methods	Substitute with filtering and standardisation methods	-Reminiscence-based and model-built collective filtering -ML
Trajectory built methods	Regulator	Has ideal condition and no parallel competence	Working on indefinite leap sites reduces the likelihood of early blending and being locked in a local ideal condition.	-Experiential search -Practiced systems -Pronouncement-making algorithms
Kemel based approaches	Control and maintenance	Probabilistic output and long training time	Predictability with statistics to analyse the occurrence of post-positive and ineffective events, and solves exercise sets locally to minimalize the exercise time	-Regression algorithms -Neural networks and their hybrid approaches -ML -Proficient systems
Randomized Probabilistic approach	Maintenance	Multidimensional computations and probabilistic yield with arbitrary variables	Uses representative reasoning to solve compound computations	-Logical neural Networks -Decision trees
Population-based methods	Design regulator and upkeep	Complex application approach, less merging speed	Attains pre-exercise with a pretty small learning rates to achieve fast convergence	-ML -Heuristic search -Expert systems

2.1 Machine learning

Machine learning (ML) consists of strategies that allow systems to accomplish a task automatically from experience (data stored in a record) without human intervention [47]. The ML implementation process starts with raw data, and progresses through feature extraction, training and evaluation, and model distribution. The process is often initiated by selecting time-series data (for example, stock and /or return data) as depicted in figure 4, and the relevant information over a given time to achieve a target. To handle complex problems, including medicinal, economic, ecological, advertising, security, and manufacturing applications. ML is commonly preferable over conventional techniques which fail or are unfavourable to multi-task. Given that there is no complex interplay between outputs and inputs, ML is the best choice for problems that have multi-parameters including the following.

- Immense data intellect: This is a data-based and information-based AI approach for data management.
- Cross-media sensing and computation: Asynchronous sensing and cognition engines that outperform humans.
- Swarm intelligence: processing information related to group behaviour
- Hybrid and improved intelligence: Integrated application of human and AI.
- Autonomous coordination and control: Operating machines and systems automatically without human interference.
- Optimized decision-making: Obtaining setpoints to maximise the efficiency of a system without any variabilities.
- Brain-inspired intelligence computation: Research principles and techniques on brain-stimulated sensing and studying,
- Considerable intelligent computation: Models and systems depend on the application of quantum calculations.

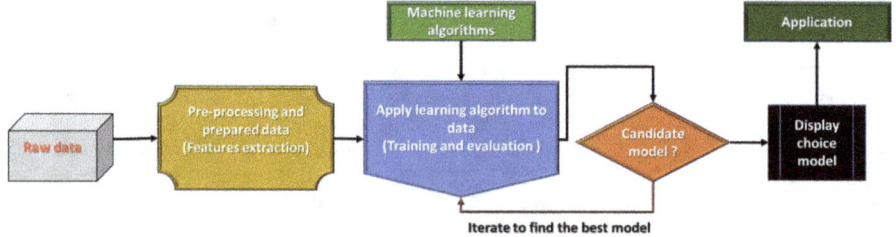

Figure 4. Machine learning workflow adapted from Mellit and Kalogirou [2]

The four classes of ML algorithms, presented in Figure 5, are supervised, unsupervised, semi-supervised, and reinforcement learning which furthermore describes the most common methods and viable techniques of ML.

i. Vector support machines, k-nearest neighbours, linear regression, logistic regression, decision trees, naive Bayes, neural networks, and random forests are examples of the supervised class of machine learning algorithms. It's unclear whether supervised learning algorithms can establish links between input and output attributes before predicting output value [47]. This makes it easy to handle regression and classification concerns with supervised learning because the inputted data are usually labelled [38].

ii. The second classification of algorithms include k-means, fuzzy means, and hierarchical clustering techniques. These are unsupervised learning algorithms trained with unlabelled data and expected outputs, hence they follow rules and patterns of the available database before being able to understand the actual data [40, 47]. This algorithm is very common in handling groupings and associations, among other things [38].

iii. Generic, graph-based and other models fall under the third classification of ML. These are semi-supervised algorithms with databases containing both labelled and unlabelled data. And this is due to the high cost of labelling and the high level of expertise required for labelling [47]. They usually involve training an algorithm to discover a model, and the most common issues that can be solved with it are classification and clustering [38].

iv. Reinforcement learning is the fourth category, such as Q-learning, deep q networks, Markov decision algorithms, etc. For presentation and analysis purposes, this sort of data is mostly utilized to transform high-dimensional data into lower-dimensional data [38]. No training data sets are required because the algorithm continuously iteratively learns from its environment. Problems involving categorization and control can be solved using this type of ML technique [47].

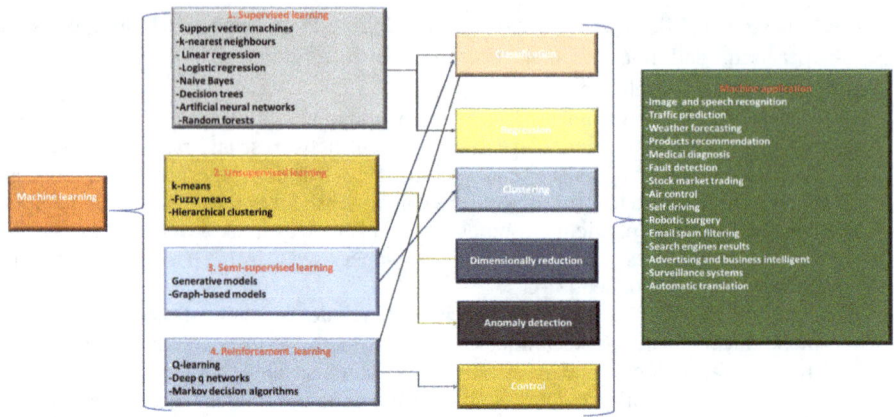

Figure 5. Machine learning algorithms and applications adapted from Mellit and Kalogirou [38] and Shehab, et al. [47]

2.2 Deep learning

Deep learning, also known as structured or hierarchical learning, has gained momentum in ML research since 2006 [46]. Herein, recent DL research has improved many aspects of signal and information processing, including ML and AI. On other hand, DL is an emblematic neural network-based approach, in that it can overfit and reduce gradients[37, 46]. This approach can extract features from massive data sets, allowing for modelling flexibility in network designs and model parameters [46].

In general, DL algorithms combine and train a set of classifiers, and then aggregate their predictions into a single prediction/decision [46, 47]. Models of DL are made up of multiple levels or stages of nonlinear processing information, such as supervised or unsupervised learning of feature extraction at progressively improved levels with increasingly arbitrary layers [37, 46, 47]. In addition, DL is an interdisciplinary field integrating studies of neural networks, AI, pattern recognition, signal processing, graphical modelling and optimization [46]. The rising processing capacity of chips (such as general-purpose graphics processing units - GPGPUs), the expanding number of training data, and current breakthroughs in ML and signal/information processing research are all important areas of DL that are currently gaining prominence [37, 43, 46].

Since then, DL algorithms are utilizing complicated compositional nonlinear functions, and dealing with labelled data. Some of the DL techniques include deep convolutional neural networks (CNN), long short-term memory (LSTM), deep belief networks, generative adversarial networks (GAN), and hybrid combinations [37, 40, 46, 47]. In this context, the most used DL in renewable energy defect detection and diagnostics as are shown in Table 3. CNN, LSTM, and GAN are among them.

Table 3. Deep learning (DL) application in power systems fault detection and diagnosis adapted from [38]

DL-based method	Diagnosis classification	Fault types	Photovoltaic (PV) systems	Complexity and software	Datasets	Real-time verification
CNN	Detection	Defect on solar cells	PV module	Medium and open source (Python)	Infrared images	online
CNN	Classification	Line-to-line & open circuit	PV string	Medium and open source (Python)	2D current and voltage signal	virtual
LSTM	Classification	Line-to-line, Hot spot, defect on solar cells	PV array	Medium and open source	6067 samples	Offline
GAN	Identification	Arc faults	PV string level	Relative complex and open source (Python)	30 000 samples	Offline
CNN	Detection	Anomalies	Large-scale PV system	Building complex and strict testing with wide open-source code	0732 images	Offline

2.2.1 Convolutional neural networks (CNNs)

The CNNs are a type of DL with a topology that looks like a grid. These are NNs that replace matrix multiplication with convolution in at least one of their layers [37, 46]. As shown in Table 3 above, software like Python is commonly used to make machine DL CNN applications for large PV facilities or farms, as well as utility grids. It comprises input and output layers with several concealed layers [46]. Figure 6 illustrates a complex CNN structure, including convolution and pooling layers in most networks. A CNN convolves the entire image as well as the intermediate layers prior to the output layer using various kernels in the convolutional layers. Image representation and network parameter sizes are reduced by using pooling layers. The calculations of pooling as well as convolutional layers are language-independent [46]. Furthermore, CNN involves two training techniques namely (i) forward stage and (ii) backward stage[46, 47]. The forward stage represents the input image with the existing parameters (weights and bias) whereas the backward step computes each parameter's gradient using chain rules. The gradients update all parameters, preparing them for the next forward computation. The network learning can be stopped after enough forward and backward iterations [40, 46, 47].

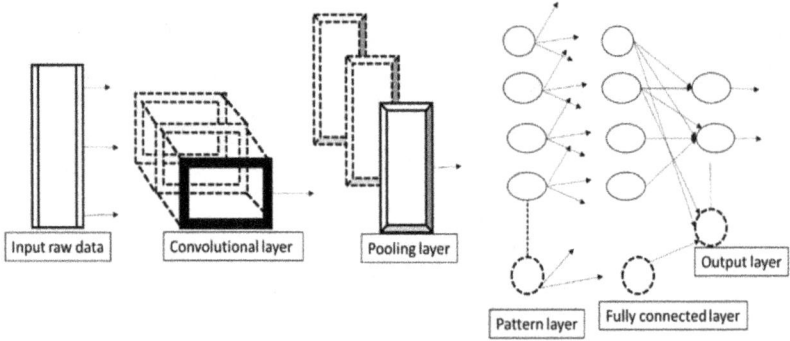

Figure 6. Convolutional Neural Networks (CNN) structure adapted from Guo, et al. [46]

2.2.2 Long short-term memory (LSTM)

The LSTM is a neural network architecture designed to process time-dependent variables in time series data sets. This type of network can use prior knowledge to forecast the future state of the variable, which is useful when the input data are not independent [46]. The Recurrent Neural Network (RNN) of LSTM uses an unrolled loop cell, which allows prior information to flow into the prediction of a subsequent step [37, 46]. Despite that, its developed form can prevent it from effectively processing long-term dependencies, as the learning process removes gradients during the back-propagation phase [46, 47]. To overcome this challenge, LSTM networks use a simple three-gate structure: input, output, and forget gates as depicted in Figure 7. This combined with a continuous gradient computation structure ensures memory retention. In general, without specific hardware and software accelerations, the computing time of LSTM can be estimated to be directly proportional to the number of parameters [38, 40, 46]. The LSTM architecture can also be utilized to display solar irradiance [41, 46]. In essence, comparing LSTM to the usual DL-based approach, LSTM network models can enhance both the predictability and efficiency of a system understudy [46].

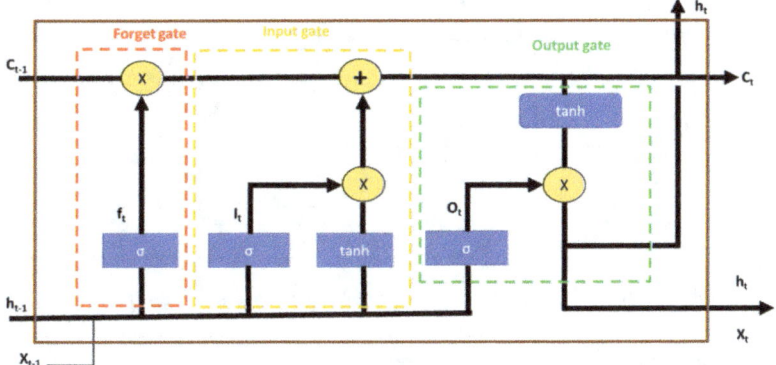

Figure 7. LSTM neural network framework adapted from [38]

2.2.3 Generative adversarial network (GAN)

To circumvent the small sample problem, generative adversarial networks (GANs) have been introduced [46, 47]. Figure 8 depicts a general structure of GANs with some key components like generator and discriminator. To achieve the best possible matching between produced and real images, GAN optimizes the discriminator and generator in parallel. In this instance, the generator creates samples with the most realistic distributions, whereas, the discriminator produces the most accurate classification result [46]. High-quality sampling and improved training stability are two common goals of GAN-based algorithms. The main goals of many GAN-based approaches are to increase performance, generate high-quality samples and improve training stability [38, 40, 46, 47]. The generator's structure can also be modified by using an online-output model or building a Laplacian pyramid framework [38, 46, 47]. In addition, multi-discriminator GAN frameworks [8, 10] are designed to stabilize the adversarial training process of GANs. Some of the heuristic approaches used to improve the training stability of GAN include virtual batch normalizing, feature matching and one-side label smoothing [44, 46].

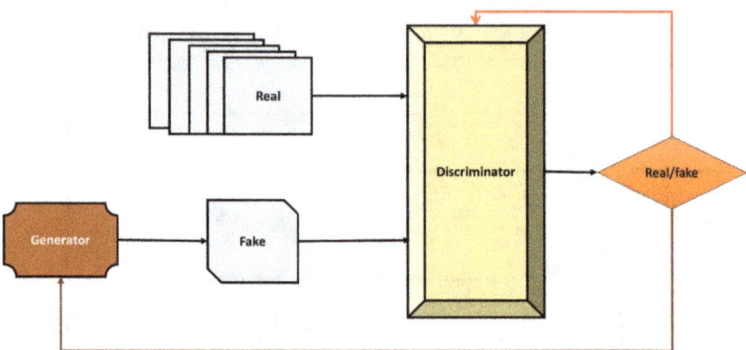

Figure 8. Generative adversarial network (GAN) structure adapted from Feng, et al. [45]

3. Application of AI in solar PV system

Most research developments on solar energy optimization have been conducted recently. Due to the uncertain aspects of PV materials, there are difficulties with the stability of the overall system [48]. Previous studies have presented more suitable optimisation strategies than conventional types [48, 49]. In terms of sizing, load demand and power generation, the optimization algorithms have shown great results in solar PV applications. Furthermore, the optimizations aid in lowering operational costs and power losses, as well as improving peak power incorporation and how best the process could be controlled.

3.1 Monitoring of PV systems

The surveillance system for PV plants collects and analyzes a variety of parameters to survey and/or examine their activity. An effective monitoring system is required to enhance the reliability and stability of any PV facility [35]. The monitoring system also maintains track of various electrical generating indicators and faults. Existing PV monitoring systems are only suitable for large-scale PV projects due to their high cost and complexity. Different components of PV surveillance systems have been documented in literature over the last decade. This covers a thorough examination of all of the primary PV monitoring evaluation methodologies and their relative performance. Sensors and their operating guidelines, and controllers utilized in data collection, transmission, storage and analysis mechanisms are all major parts of PV monitoring systems. All these factors need to be understood in order to design viable, cost-effective, and efficient PV monitoring systems for small- and medium-sized PV systems without sacrificing required efficiency. PV monitoring systems are designed to ensure data availability on the energy harvesting operating temperature analysis, and energy loss connected with various defects that may occur. The observed data can then be utilized for early identification, climate change evaluation, and other purposes [38, 50].

As a result, lots of efforts have gone into designing efficient PV monitoring systems. The monitoring system would include various commercial goods that have been integrated into it. With the increased proliferation of commercial products based on numerous concepts, it is critical to study each one's functioning and characteristics. For a viable PV monitoring system, selecting the right product for a specific climatic situation is critical. Various elements of PV monitoring systems have been reported throughout the last decade. Figure 9 depicts a general block structure of a PV monitoring system classification. Owing to the lack of sensors, space-based systems may be cost-effective in this regard. The drawback of space-based systems is their poor precision, which is heavily influenced by weather situations, which is undesired. As a result, the extent of this analysis of PV monitoring systems is limited to ground-dependent systems.

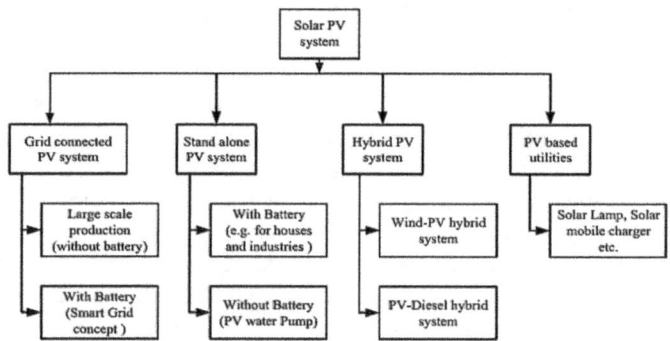

Figure 9. A block representation of the PV systems classification adapted from Mellit [33].

3.2 PV fault detection and diagnosis (FDD) methods

If issues in parts (modules, interconnectors, converters or inverters) of PV systems (hybrid PV, stand-alone, or grid-connected systems) are not quickly identified and rectified, they can have a substantial effect on the performance and energy output of a PV system [51]. Moreover, some faults (e.g. arc, ground and line faults) can aggravate the threat of fire-outbreak. The FDD systems are critical for PV system stability, performance and security. When a PVM fails, the problem is typically linked to the system warranty. PVS flaws result in unanticipated safety concerns, lower performance, power accessibility, system reliability, and safety [48].

Many researchers have identified and investigated PVM faults of various forms. Some of these problems are discolouration, cracking, degradation of the anti-reflective coating, blistering, staining, oxidation and corrosion of the bus bars, split encapsulations over cells and joints and loss of backside adhesion [26, 33]. Encapsulation, module corrosion, cell cracking, and PV inverter are some of the failure modes that are commonly occurring. In general, PVM defects can be divided into two categories: permanent and temporary. Permanent flaws include yellowing of cells, scrapes, delamination, charred cells and bubbles. So, by simply replacing the damaged modules, such defects can be avoided. However, temporal issues caused by dust build-up, snow, partial shading impacts and dirt on PVM can be eliminated without changing the damaged PVM. Furthermore, the failure may originate from internal or external sources, and the combined scenario could lead to a reduction in the power output, performance, and dependability of the PVS.

3.3 Employment of AI technologies for sizing PV systems

The climatic conditions, such as temperature, direct sunlight, and dust, have an impact on PV cells designed for outdoor use and their ability to generate power [48]. Each single PV cell is of the order of a few inches and a single PV panel consists of a large number of PV modules. Many cells are combined to form a module and various modules are connected in batches to generate additional power. As a result, a vast system built on local approaches has been employed to optimize operational parameters to enhance the coherence of solar PV systems. Because the purpose of optimization is to increase output while lowering costs, it is necessary to evaluate the benefits and drawbacks of the particular system. Researchers have begun creating purposeful approaches by developing models that will maximize the benefits while minimizing the drawbacks [33]. Many studies have addressed conventionally optimized techniques to solve challenges connected to orientation, operation, and renewable energy systems. Creating procedures are influenced by animal colonies and the mechanisms of biological or physical operations. Furthermore, intelligent techniques are sophisticated and time-consuming [29]. To precisely evaluate the power output of present operating parameters, PV systems require consistent and valid performance data. The four compartments of a solar system are batteries, solar cells, load, and inverters. The components of a solar system must be selected based on size, cost, and use and thus, crucial to look at the system design in terms of energy generation capacity, economics, and reliability.

3.4 Modelling of a solar PV generator

Solar PV power facilities that are connected to the grid are becoming more widespread in India. The majority of these have power outputs ranging from a few hundred kilowatts to tens of megawatts [52]. Small grid-connected solar power plants of 1 or 2 KWs are not being deployed in India, like in many other countries. The ability of solar PV power plants to generate electricity has increased dramatically during the previous several years from a universally installed output of 7 GW reported in 2006 to 70 GW in 2011 with a ten-fold growth in five years. So far, six countries have now added more than 1 GW of capacity this year. For solar PV power generation, this being the most frequently used mode of operation was regarded as the only mode of operation for PV systems till a few decades ago. The power plant served as an electric power generator for local loads in this case. Because solar power generation is restricted by the amount of sunshine available, energy storage technologies in battery form are commonly used [30].

Some systems, such as water pumping systems, do not require any energy storage. This category includes large-scale solar PV power facilities installed around the world. These power systems are linked to the grid which produces and feeds electricity into it whenever sunshine and the grid are available. This type of solar power plant has been erected with a capacity of up to 214 MW all over the world. These massive power facilities lack energy storage and feed power into the grid at high-tension voltage levels. Energy storage is needed to support energy generation during periods of low solar irradiation and to keep the system stable when it is running in grid-disconnect mode. The inability to manage the source of power is disadvantageous to RE sources like sun and wind. Weather and cloud conditions are beyond our control due to unpredictable variations in sunlight and wind [36]. As a result, loads cannot rely only on these sources to function. This battery-storage architecture is also particularly important in India's grid conditions, which are prone to frequent outages. The presence of batteries during outages has two advantages as it prevents the loss of solar energy generation due to grid failures (uninterrupted generation), and provides consistent electrical energy to local loads (UPS function).

3.5 Solar water heating systems (SWHs)

SWHs capture sunlight to convert into heat [53]. The temperature at which solar thermal collectors effectively deliver heat can be classified as low-temperature collectors (unglazed mats for water heating), mid-temperature collectors (glazed and insulated collectors) and high-temperature collectors (evacuated tubes and the focusing collectors). Water heating accounts for 4.5 quad/year of 37.6 quads/year of the total United States building energy usage. The distribution of energy for the commercial and residential sites consumption is represented in Figure 10.

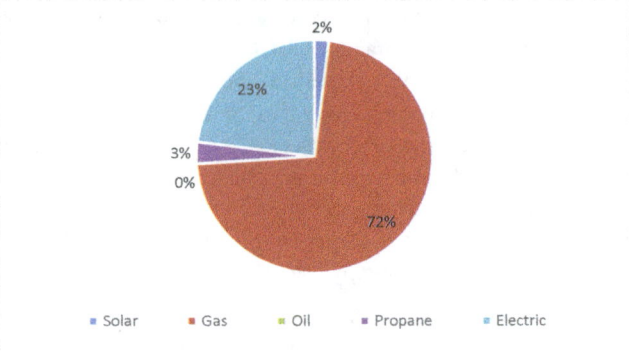

Figure 10. Commercial and residential energy site consumption

SWHs have been categorized into two sub-groups as active and passive which all work in either a direct or indirect manner. Active SWH systems have better efficiency than passive systems, with values ranging from 35 % to 80 % which are usually more difficult to achieve [29]. As a result, they are best suited to industrial applications with higher load demands and applications where the collector and service water storage tanks do not need to be adjacent to one another. Passive systems are less capital intensive, and easy to design and install. These are best suited to home applications with a medium load requirement. In general, further research and developmental efforts are required to motivate the present levels of efficacy of traditional hot water production methods. They tend to solve the drawbacks of a typical SWH system which includes corrosion in pipes due to water usage and reduced thermal transmission [28].

The use of a solar water heating system will aid in the reduction of harmful emissions produced by fossil fuels. Solar collectors, storage tanks, and interconnecting pipework are the fundamental components of a solar water heating system. Solar collectors are commonly positioned on the top of a structure to capture solar energy and transport fluid, such as water or a heat transfer liquid with a boiling point. The fluid absorbs and transmits heat from the solar collectors to the incoming water. The solar collectors, flat-plate collectors and evacuated tube collectors (ETC)) are often utilized in the solar thermal system rectangular box with a metal absorber plate within the flat-plate collector. The box is painted in a dark tone to improve solar energy absorption. The evacuated tube collector is made up of parallel rows of translucent glass tubes that look like fluorescent tubes and contain an inner and outer glass tube with a specific coating that absorbs solar energy. The ETC is often more expensive than the other two types of solar collectors due to a more complicated production process. On the other hand, the ETC is considered more efficient per unit area than the flat plate collector (FPC), requiring less installation.

4. Challenges of effective AI application in solar PV system

4.1 Solar energy optimization

Solar energy is among the commonly used RES, and it leads the market for renewable energy. However, despite the significant progress that has occurred in recent years, solar energy is confronted with significant challenges that could stifle this growth. These drawbacks related to techniques, administration, finance, and reliability [54]. In contrast, optimization helps to mitigate the downsides of solar energy systems while also increasing their dependability. Consequently, improved solar energy optimization can contribute to alleviating the unpredictability associated with energy generation [55]. A significant amount of money is being invested in the photovoltaic (PV) power technology sector to increase performance. The downsides of PV cells include the fact that they stop producing power when there is no solar radiation shining on them and that their effectiveness is low compared to other forms of energy generation [56]. This could result in a failure to meet the capital investment necessary to make the technology lucrative. Subsequently, solar energy storage devices have been suggested among the options for overcoming the lack of light and flattening the electricity production and demand curves. This technique depends on batteries, which are sometimes big, enormous, and heavy, occupying a significant amount of space, and need constant upkeep or even substitution now and then [57]. When compared to the past decades, the majority of these systems' components are now available at reasonable cost due to continued technological advancements such as greater PV conversion efficiency of solar-electricity-generating solar panels [4].

RESs can provide a variety of technological advantages to the electrical power system, including increased voltage profile consistency, reduced power losses, and lowering of energy tariffs [58]. RESs can also increase the sustainability of the electrical power system. Capital expenses, economic concerns, market constraints, public recognition, stakeholders, legislation, and policies are some of the non-technical barriers that must be overcome. The PV system optimization procedures necessitate the use of following critical inputs [4]:

- To achieve a preferred optimization, it is critical to obtain correct data for the primary characteristics of the solar system, which include wind speed, ambient temperature, dust, humidity, and sunshine. However, the hourly or daily collection of these data remains one of the most difficult tasks in optimisation, despite the existence of prediction techniques such as ANN.

- Load predicting is necessary to have access to the outline of a maximum loading requirement over one year to determine the optimal PV device size.
- Validity of models: When optimising, it is critical to have a model that is as precise as possible. To be accurate, the model must take into account all of the important aspects that influence system effectiveness and performance.
- The wide range of models: In terms of transformation efficiency, it is quite challenging to find identical PV models. There are numerous PV brands commercially available, and significant developments in manufacturing activity have resulted in conversion efficiencies that are similar to identical models.
- Accessibility and application of the offered techniques: Optimization should produce lucid and exact outcomes. This can be accomplished by merging two methodologies, such as analytical techniques and artificial intelligence technology.

Among the essential aspects, often used to characterise PV system efficiency are the peak power Pmax, short-circuit current density Jsc, the fill factor (FF) and open-circuit voltage Voc, which are all measured in watts. A study by Meral and Dincer [59] identified the major parameters that may influence the efficiency of a solar power production facility. Researchers are also confronted with the most current trends in effective improvement in solar energy utilization, and pertinent investigation issues. It is essential to analyse the pros and cons of the different facilities that are under review because the objective of the optimization is to enhance performance while simultaneously reducing overall cost. Many conventional optimization approaches, as well as recent heuristic techniques, have been used extensively in numerous research to resolve issues associated with the development, functionality, and mechanism of RES, and this is reflected in the results of many investigations. A major contributor to the formation of the most intelligent ways has been the collaborative behaviour of animal colonies, and a secondary contributor has been the intelligence process of biological and physical activities. Furthermore, intelligent techniques are both complicated and time-consuming [4]. Recent advances in computation techniques have made it possible to use intelligent techniques and optimization algorithms in the area of solar energy production. The optimization approaches can be classified as cost optimization, size optimization, and control strategy optimization [60]. It is critical to have consistent and efficient performance data for PV facilities to properly calculate the power produced and the capacity of the system under existing operational parameters. This dependable data assists in making the best decision possible in the field of performance and handling. However, the optimization and performance of solar PV could be achieved via the investigation of the different factors that influence the performance of the power system and the exploration of various phases that can be employed to optimise the efficiency of the power plant on a larger scale [61].

4.2 PV-dependent hybrid facility optimization

One of the most significant drawbacks linked with the usage of REs is their unpredictable character and failure to function effectively as a result of inconsistent and unaccepted fluctuating nature, which frequently results in over-sizing and a rise in capital expenses. These challenges have lately been addressed by the hybrid method's structures, which have gained widespread recognition as a viable remedy to these issues [62]. The goals of a hybrid renewable energy system (HRES) include lowering the overall expenditure of the facility, minimising the energy storage capacity, improving performance, and increasing dependence while maintaining or improving stability [63]. Many studies have focused on the application of conventional optimization methods and more recent heuristic methods to tackle challenges connected to the design and operation procedures of HRES [64]. Biological or physical intelligence mechanisms or animal colonial behaviours are the foundations of the vast majority of optimization strategies developed [65]. Though data-driven optimization techniques are limited by sophisticated computation and intensive training periods, current development in advanced computer techniques, large storage volumes, and incredibly fast computing capabilities have alleviated the computational sophistication difficulties [66, 67]. For instance, depending on multiple mixes of independent producing systems, one of the optimization models employed in hybrid systems estimates the ideal design of HRES. Using this model, the optimal configuration of HRES has been determined [68]. A further study by the authors on the optimization of system size revealed that off-grid hybrid facilities, which include biomass and solar PV systems, may give a practicable option. Khatib and colleagues [69] developed an approach to assist in complementing predictions amid solar power and small hydro power systems. An optimisation algorithm that linked hydrology with solar irradiance data was used to build the methodology. The programme then advised feasible adjustments in the system design that may boost their competency in the field.

A multi-purpose genetic algorithm was applied to optimise the design of a PV-wind hybrid facility in a study by Eren et al., [70] and the hybrid system was improved as a result. The optimization techniques made use of a variety of methods, including stochastic processes, intuitive strategies, deterministic processes, and numerical processes. The stochastic process was used to deal with solar energy unpredictability, which included the hourly solar energy series. The proposed study could result in a sustainable energy supply with the least amount of investment. The concept was to determine the most efficient way to expand or improve the efficiency of the renewable energy system that had been implemented.

Optimization of the system's influential factors may be necessary to get the desired outcomes.

4.3 External factors of solar energy generation

The performance of PV cells designed for outdoor use, as well as their ability to generate energy, is influenced by the weather conditions in the area, such as temperature, direct sunshine, and dust [71]. A PV panel can consist of several PV modules, and the PV cells that make up each module are typically just an inch wide in diameter. To create additional power, several cells are linked together to create a module, and numerous modules are connected to produce arrays. As a result, a large-scale system is built. The optimization of process variables to increase the performance of solar PV systems rely on both conventional and intelligent techniques. These factors (temperature, sunlight, humidity, and dust) determine the output power produced by a PV power system. Each of these parameters changes with the passage of time and the relevance of the issue. While it is true that these external factors cannot be managed directly, their impacts can be mitigated to a certain extent. As a result, it is necessary to examine the effect of external conditions on system efficiency and to propose relevant interventions to maximise system efficiency. The open-circuit voltage and short-circuit current of a PV cell are the two most significant features that should be examined to demonstrate the disparity in the supply of power that is produced by the cell. The rise in ambient temperature brings a reduction in the voltage of the open circuit, as well as a slight increase in the current of the short circuit, which results in a decrease in efficiency. The impact of temperature on the operation of solar cells was investigated with the help of MATLAB [72]. Increasing temperature caused a rise in reverse saturation current and a decrease in Voc, which caused a fall in the fill factor and, thus, a decrease in the solar cell capacity, as per the simulation study. Meanwhile, a drop in bandgap was associated with a rise in Jsc, which led to an overall improvement in cell efficacy. As a result of the variation of Voc and Jsc, the efficiency of the solar system decreased. Chander et al. [73] conducted an investigation and revealed that temperature was a significant factor influencing the efficiency of photovoltaic cells. After the surface temperature climbed above 298 K, the highest generated power decreased by around 0.47 % for every 274 K rise in temperature [74].

Al-Showany [75] investigated the impact of certain environmental variables, including dust and ambient temperature, on the efficiency of two similar photovoltaic cells, with regards to the voltage generated and electrical power output by the cells. The researcher used a water-cooling technique to reduce the PV cells' temperature and enhance the amount of solar irradiation that reached the cells. According to numerous studies, the temperature is the major parameter influencing the performance of a PV cell [76]. Consequently, lowering the surface operating temperature of a PV module is a highly effective method of increasing the efficiency of the module. The amount of heat energy retained in the PV cells when the system is in service can be minimized by cooling the module. Another important aspect to address is dust, which obscures part of the sun rays and reduces the performance of the PV system [77]. The authors created a timetable for the recognition, maintenance, and cleaning routine for photovoltaic facilities as a mitigative measures to maximise the performance of the PV system.

4.4   Challenges in the development of solar energy systems

Solar cells, batteries, an inverter, and a load are the four essential parts of a solar-powered energy facility. The components of a solar energy system must be selected on the basis of their size, price, and intended use. It is important to evaluate the development of the system with regards to energy production capacity, economic viewpoint, and dependability before proceeding. The scientists are primarily concerned with optimising the system architecture by selecting the most appropriate components. To boost the performance of PV, Stritih [78] took into account superficial temperatures and the PCM cooling system. PCM improved the reliability of efficiency by 14 %, according to the results of the experiment. Zheng and colleagues [79], as well as Dufo-Lopez and colleagues [80], have created a stochastic-heuristic approach that takes into account wind speed, solar irradiance, and temperature to minimise the cost of operation. Motaleb and colleagues [81] used solar energy and a battery to develop a stochastic process to lower the cost of produced energy as much as possible. To minimise energy loss, Hashemi-Dezaki et al. [82] looked into wind and solar energy sources. In their study, Dufo-López et al. [83] attempted to reduce net cost by considering solar irradiation and load. Alternatively, affordable reflecting components such as lenses or mirrors were employed to describe the performance of the PV facility. In addition, Maximum Power Point Tracking (MPPT) is a sophisticated approach to maximise the amount of energy generated by a PV module and consequently improve the conversion power of the system [84].

Several PV module designs are not capable of providing the same output under similar working parameters, so the modules must be properly installed in the same system under the same operating settings. A variety of technical components, like batteries, chargers, electrical power devices, and cabling, were considered during designing of the system, as these components are critical to the functioning and performance of PV facilities. It was advised that bigger capacity batteries be used in PV systems since the batteries are charged and discharged on a relatively regular basis. Several types of batteries are currently available, including flooded and valve-regulated. Compared to valve-regulated batteries, flooded batteries are more durable but need regular maintenance. The inverters in different designs are important solar systems but all are not good for solar power. It is critical to adopt well-developed circuit techniques to minimize the conductivity and switching losses of semiconductor components, and the power dissipation generated by the connectivity of cables, to improve power quality. In reality, numerous studies suggest that the performance of PV systems can be increased by improving the output power of the systems' solar panels. Many ways are employed in this vein to improve the regulating aspect and efficiency of the PV cell. These include but are not limited to:

- The quality of the base component is being improved to capture more radiation.
- Including substitute, RES and optimising the size of the system to maximise its economic viability is a priority.
- To maximise output power, external parameters such as temperature, irradiation, and dust must be controlled properly.

4.5 Solar energy transformation

A solar cell, inverter, and a solar charge controller make up the components of a solar photovoltaic (PV) system. By capturing more radiation, an improved manufacturing process for solar cells that incorporates new treatments could boost the transformation OPERATION of the solar cells. The major elements of the inverter, including the power switches and magnetic parts, should be carefully chosen to achieve optimal efficiency [85]. Utilizing filters, the solar charge controller (also known as a DC controller) can increase the performance of the system [86, 87]. It is necessary to reduce the harmonic content of the output from a DC-DC converter using filter circuits. Passive filters, LLCC, LCL, and LC have been utilised to reduce the harmonic disturbances and improve the power quality of a system. Moreover, filter capacitors were employed to minimise increased current ripples [88, 89]. The intrinsic control algorithm of the pulse width modulation (PWM) maximises the amount of energy conveyed to the battery bank and, as a result, enhances the battery bank's life span [90]. Furthermore, since inverters play such an important function, they must be properly maintained. Otherwise, the user may be unable to obtain any service, even though the PV system is producing electricity [91].

It is necessary to start with the solar cell module and the phase decision to build an effective solar power system. This technique is responsible for converting the energy of photons into electricity. After being attacked by photons, they gain energy and generate a potential difference in the circuit, which in turn produces a voltage that is required for the operation of the circuit. Semiconductor components including silicon (Si) are frequently utilised in the manufacture of photovoltaic (PV) cells. Silicon nanowires (SiNW) contribute actively to boosting the performance of solar cells, since the length enhancement in SiNW increases the anti-reflection characteristic of the solar cell [92]. However, two-step thermal oxidation of multi-crystalline carriers can minimize surface recombination of the carrier and increase cell performance [93]. In the spectral spectrum, this layer boosts light absorption approximately by 4 % and this results into a rise in the overall productivity of solar cells. Furthermore, the design of photovoltaic (PV) modules that captures a certain wavelength of sunlight from solar radiation has the potential to improve performance [94]. Several ways to enhance the operation of the solar system, including optical supervision, were discussed by Zhang and Toudert [95] in their review paper. The following are some examples of different techniques.

- Antireflective coatings for the cell's uses.
- Modification in the vertical layout of the cell
- Incorporation of plasma or dielectric nanostructures hooked on various cell layers.
- The framework of the cell's inner interfaces at the wavelength scale.

5. Prospects and future work consideration

The implementation of AI in smart energy systems has different types of bottleneck challenges, which include poor data standard, unavailability of data, tuning AI network factors, technological facility issues, very limited skilled professionals, risk management and regulatory compliance issues [41]. The detection and identification of faults in developing energy systems are among some complex challenges. Herein, some studies have concluded that data security and information scarcity are two of the most significant drawbacks confronting the energy sector today [37]. Moreso, substandard sensors, and controlled devices that are employed in the operation of energy systems and data evaluation reduce the system's dependability and efficiency. The power grid's time-consuming connections and robust integration, as well as its high data dimensionality, pose significant difficulties in the energy market [37, 41, 43, 44]. Using AI to incorporate RES (solar and wind) is a challenging task for grid operation computing, which is among the most effective supercomputers. Also, quantum techniques facilitate AI-based machine learning procedures and improve system processing capacity [39, 46]. ML and DL can improve fault identification, categorization, and positioning of surveillance systems [47]. The most efficient strategy to avoid and reduce solar plant failures is to incorporate a fixed fault identification system based on a smart tool.

Infrared and fuzzy logic image-based fault detection/identification, localization, and categorization are costly to implement. Using IoT, a data-acquisition circuit and Wi-Fi module can send data to the cloud [39, 40, 46]. The present condition of the PV system can be put on a website, with the information provided on the fault type and the faulty module or string [41, 46]. Moreover, operators are alerted via SMS if a problem arises. In addition, the application of AI in the digital transformation of power systems has been cited as having substantial potential for enhancing power system network stability and facilitating crucial changes [37, 44]. As shown in Figure 11, AI is being used to implement the design, forecasting, control, optimization, maintenance and cyber security components of the power system [44]. This suggests that the effective development of AI is a promising sustainable strategy to mitigate carbon footprint as direct rebound affects energy production.

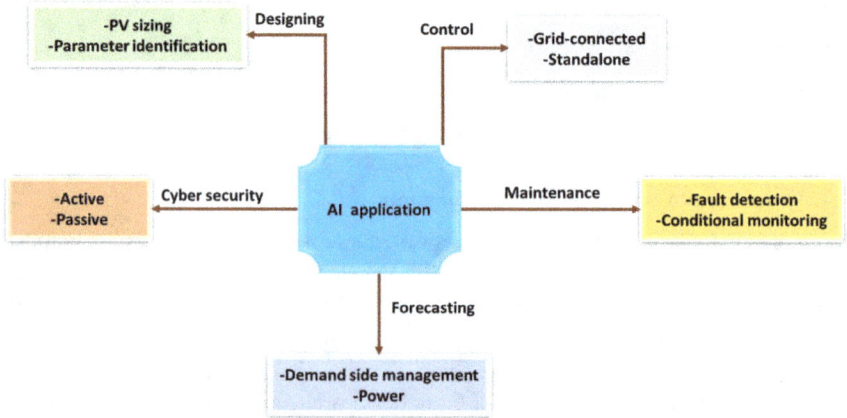

Figure 11. Schematic diagram of artificial intelligence (AI) applications in power systems adapted from Kurukuru et al. [44]

Conclusion

Artificial intelligence approaches DL and ML, have lately acquired prominence in this subject, attracting several researchers who are working to design and apply new defect detection and diagnostic techniques. In light of the availability of vast quantities of data and the accessibility of supercomputers, it is reasonable to predict rapid progress in the employment of ML and DL techniques in this field in the not-to-distant future. As a result, developing intelligent problem diagnostic systems that are dependent on AI and IoT are hoped to become increasingly important. DL is a subset of ML, and it is becoming increasingly popular. Machine learning makes use of mathematical algorithms that enable machines to understand on their own, and algorithms can learn from data by extracting insights from the data. Deep learning applies neural networks and can learn from raw data without feature extraction, which is the primary distinction between ML and DL. Deep learning is also more expensive than machine learning.

ML and DL have promising prospects to significantly improve the validity and efficiency of surveillance systems used to diagnose, identify, characterise, and localise various kinds of errors in a PV energy system, as well as to improve their overall performance. Apart from utilising the current security and diagnosis systems, the most efficient method to safeguard and prevent faults in a photovoltaic facility is to incorporate an integrated fault diagnosis technique that is based on advanced devices into the system.

Finally, readers are presented with comprehensive difficulties, recommendations, and perspectives to provide a vivid insight into future developments in the field, such as areas that require improvement and further study. A clear understanding of the precise implementation of ML, DL, and IoT approaches, as well as the discrete challenges encountered in this extremely notable and prospective field of research, can be gained from this study, which can benefit researchers in both academic institutions and industrial sectors.

References

[1] L. Bastida, J. J. Cohen, A. Kollmann, A. Moya, J. Reichl, Exploring the role of ICT on household behavioural energy efficiency to mitigate global warming, Renewable and Sustainable Energy Reviews. 103 (2019) 455-462. https://doi.org/10.1016/j.rser.2019.01.004

[2] A.Q.A. Shetwi, M. Hannan, K.P. Jern, M. Mansur, T. Mahlia, Grid-connected renewable energy sources: Review of the recent integration requirements and control methods, Journal of Cleaner Production. 253 (2020) 119831. https://doi.org/10.1016/j.jclepro.2019.119831

[3] I. IRENA, Renewable power generation costs in 2017, Report, International Renewable Energy Agency, Abu Dhabi, 2018.

[4] O.A.A. Shahri, F.B. Ismail, M.A. Hannan, M.S.H. Lipu, A.Q.A. Shetwi, R.A. Begum, N.F.O.A. Muhsen, E. Soujeri, Solar photovoltaic energy optimization methods, challenges and issues: A comprehensive review, Journal of Cleaner Production. 284 (2021) 125465. https://doi.org/10.1016/j.jclepro.2020.125465

[5] A. Harrouz, M. Abbes, I. Colak, K. Kayisli, Smart grid and renewable energy in Algeria, in IEEE 6th International Conference on Renewable Energy Research and Applications (ICRERA), 2017, pp. 1166-1171. https://doi.org/10.1109/ICRERA.2017.8191237

[6] H. Lund, B.V. Mathiesen, Energy system analysis of 100% renewable energy systems-The case of Denmark in years 2030 and 2050, Energy 34 (2009) 524-531. https://doi.org/10.1016/j.energy.2008.04.003

[7] A. Qazi, F. Hussain, N.A. Rahim, G. Hardaker, D. Alghazzawi, K. Shaban, K. Haruna, Towards sustainable energy: A systematic review of renewable energy sources, technologies, and public opinions, IEEE access. 7 (2019) 63837-63851. https://doi.org/10.1109/ACCESS.2019.2906402

[8] H.M.A. Maamary, H.A. Kazem, M.T. Chaichan, Renewable energy and GCC States energy challenges in the 21st century: A review, International Journal of Computation and Applied Sciences. 2 (2017) 11-18. https://doi.org/10.24842/1611/0018

[9] M.A. Hannan, Z.A. Ghani, M.M. Hoque, M.S.H. Lipu, A fuzzy-rule-based PV inverter controller to enhance the quality of solar power supply: Experimental test and validation, Electronics 8 (2019) 1335. https://doi.org/10.3390/electronics8111335

[10] IEA, "World Energy Outlook 2015," 2015. https://www.iea.org/reports/world-energy-outlook-2015

[11] I. E. A. IEA, "Key World Energy Statistics 2020," Paris, 2020. https://www.iea.org/reports/key-world-energy-statistics-2020/transformation

[12] M. Gul, Y. Kotak, T. Muneer, Review on recent trend of solar photovoltaic technology, Energy Exploration & Exploitation 34 (2016) 485-526. https://doi.org/10.1177/0144598716650552

[13] A. K. Raturi, "Renewables 2019 Global Status Report," 2019.

[14] A. Mellit, S.A. Kalogirou, Artificial intelligence techniques for photovoltaic applications: A review, Progress in energy and combustion science 34 (2008) 574-632. https://doi.org/10.1016/j.pecs.2008.01.001

[15] R. Belu, Artificial intelligence techniques for solar energy and photovoltaic applications, in: M.K. Pour, S. Clarke, M.E. Jennex, A.V. Anttiroiko, S. Kamel, I. Lee, J. Kisielnicki, A. Gupta, C.V. Slyke, J. Wang, V. Weerakkody (Eds.), Robotics: Concepts, methodologies, tools, and applications, IGI Global., 2014, pp. 1662-1720. https://doi.org/10.4018/978-1-4666-4607-0.ch081

[16] A.T. Lahiani, A.B.B. Abdelghani, I.S. Belkhodja, Fault detection and monitoring systems for photovoltaic installations: A review, Renewable and Sustainable Energy Reviews 82 (2018) 2680-2692. https://doi.org/10.1016/j.rser.2017.09.101

[17] A. Mellit, G.M. Tina, S.A. Kalogirou, Fault detection and diagnosis methods for photovoltaic systems: A review, Renewable and Sustainable Energy Reviews 91 (2018) 1-17. https://doi.org/10.1016/j.rser.2018.03.062

[18] S.R. Madeti, S. Singh, A comprehensive study on different types of faults and detection techniques for solar photovoltaic systems, Solar Energy 158 (2017) 161-185. https://doi.org/10.1016/j.solener.2017.08.069

[19] A.Y. Appiah, X. Zhang, B.B.K. Ayawli, F. Kyeremeh, Review and performance evaluation of photovoltaic array fault detection and diagnosis techniques, International Journal of Photoenergy 2019 (2019) 6953530. https://doi.org/10.1155/2019/6953530

[20] I.U. Khalil, A.U. Haq, Y. Mahmoud, M. Jalal, M. Aamir, K. Mehmood, Comparative analysis of photovoltaic faults and performance evaluation of its detection techniques, IEEE Access 8 (2020) 26676-26700. https://doi.org/10.1109/ACCESS.2020.2970531

[21] M.M. Rahman, J. Selvaraj, N.A. Rahim, M. Hasanuzzaman, Global modern monitoring systems for PV based power generation: A review, Renewable and Sustainable Energy Reviews 82 (2018) 4142-4158. https://doi.org/10.1016/j.rser.2017.10.111

[22] A. Hamied, A. Boubidi, N. Rouibah, W. Chine, A. Mellit, IoT-Based smart photovoltaic arrays for remote sensing and fault identification, in: M. Hatti (Eds.), Smart Energy Empowerment in Smart and Resilient Cities, Springer International Publishing, 2020, pp. 478-486. https://doi.org/10.1007/978-3-030-37207-1_51

[23] G. Wheatley, R.I. Rubel, Design improvement of a laboratory prototype for efficiency evaluation of solar thermal water heating system using phase change material (PCMs), Results in Engineering 12 (2021) 100301. https://doi.org/10.1016/j.rineng.2021.100301

[24] C.C. Kung, J.E. Mu, Prospect of China's renewable energy development from pyrolysis and biochar applications under climate change, Renewable and Sustainable Energy Reviews 114 (2019) 109343. https://doi.org/10.1016/j.rser.2019.109343

[25] M. Romero, J.G. Aguilar, E. Zarza, Concentrating solar thermal power, in: D.Y. Goswami, F. Kreith (Eds.), Energy Efficiency and Renewable Energy Handbook, CRC press, 2015, pp. 1261-1370. https://doi.org/10.1201/b18947-51

[26] A. Chatterjee, D.M. DeLorenzo, R. Carr, T.S. Moon, Bioconversion of renewable feedstocks by Rhodococcus opacus, Current opinion in biotechnology 64 (2020) 10-16. https://doi.org/10.1016/j.copbio.2019.08.013

[27] E.A. Williams, M.O. Raimi, E.I. Yarwamara, O. Modupe, Renewable energy sources for the present and future: An alternative power supply for Nigeria, Energy and Earth Science 2 (2019) 18-44. https://doi.org/10.22158/ees.v2n2p18

[28] H.M. Steinhagen, Concentrating solar thermal power, Philosophical Transactions of the Royal Society A: Mathematical, Physical and Engineering Sciences 371 (2013) 20110433. https://doi.org/10.1098/rsta.2011.0433

[29] K.M. Powell, K. Rashid, K. Ellingwood, J. Tuttle, B.D. Iverson, Hybrid concentrated solar thermal power systems: A review, Renewable and Sustainable Energy Reviews 80 (2017) 215-237. https://doi.org/10.1016/j.rser.2017.05.067

[30] M. Romero, A. Steinfeld, Concentrating solar thermal power and thermochemical fuels, Energy & Environmental Science 5 (2012) 9234-9245. https://doi.org/10.1039/c2ee21275g

[31] I. Purohit, P. Purohit, Technical and economic potential of concentrating solar thermal power generation in India, Renewable and Sustainable Energy Reviews 78 (2017) 648-667. https://doi.org/10.1016/j.rser.2017.04.059

[32] J. Li, Scaling up concentrating solar thermal technology in China, Renewable and Sustainable Energy Reviews 13 (2009) 2051-2060. https://doi.org/10.1016/j.rser.2009.01.019

[33] A. Sawner, S. Shukla, Application of "PVsyst" to simulate the 100 kWp Rooftop solar grid-tied PV System at college campus: Jabalpur Engineering College, India, 5th International Conference on Electrical, Electronics, Communication, Computer Technologies and Optimization Techniques (ICEECCOT), 2021, pp. 455-459. https://doi.org/10.1109/ICEECCOT52851.2021.9707970

[34] S. Sinha, S.S. Jasial, G. Sinha, Optimization of solar energy storage in a battery for hybrid photovoltaic system," EasyChair, 2516-2314, 2020.

[35] R. Dash, S. Swain, Battery Storage Photovoltaic Grid-Interconnected System: Part-IV," in Proceedings of the International Conference on Soft Computing Systems, 2016, Springer, pp. 799-807. https://doi.org/10.1007/978-81-322-2671-0_75

[36] K.S.S. Pundir, N. Varshney, G. Singh, Comparative study of performance of grid connected solar photovoltaic power system in IIT Roorkee campus, in International Conference on Innovative Trends in Science, Engineering and Management, New Delhi, India, 2016, pp. 422-31.

[37] F.P.G. Márquez, A.P. Gonzalo, A comprehensive review of Artificial Intelligence and wind energy, Archives of Computational Methods in Engineering 29 (2021) 2935-2958. https://doi.org/10.1007/s11831-021-09678-4

[38] A. Mellit, S. Kalogirou, Artificial intelligence and internet of things to improve efficacy of diagnosis and remote sensing of solar photovoltaic systems: Challenges, recommendations and future directions, Renewable and Sustainable Energy Reviews 143 (2021) 110889. https://doi.org/10.1016/j.rser.2021.110889

[39] T. Ahmad, D. Zhang, C. Huang, H. Zhang, N. Dai, Y. Song, H. Chen, Artificial intelligence in sustainable energy industry: Status Quo, challenges and opportunities, Journal of Cleaner Production 289 (2021) 125834. https://doi.org/10.1016/j.jclepro.2021.125834

[40] A. Mellit, S.A. Kalogirou, L. Hontoria, S. Shaari, Artificial intelligence techniques for sizing photovoltaic systems: A review, Renewable and Sustainable Energy Reviews 13 (2009) 406-419. https://doi.org/10.1016/j.rser.2008.01.006

[41] O.A.A. Shahri, F.B. Ismail, M.A. Hannan, M.S.H. Lipu, A.Q.A. Shetwi, R.A. Begum, N.F.O.A. Muhsen, E. Soujeri, Solar photovoltaic energy optimization methods, challenges and issues: A comprehensive review, Journal of Cleaner Production 284 (2021) 125465. https://doi.org/10.1016/j.jclepro.2020.125465

[42] S.K. Jha, J. Bilalovic, A. Jha, N. Patel, H. Zhang, Renewable energy: Present research and future scope of Artificial Intelligence, Renewable and Sustainable Energy Reviews 77 (2017) 297-317. https://doi.org/10.1016/j.rser.2017.04.018

[43] R. Belu, Artificial intelligence techniques for solar energy and photovoltaic applications, in: M.K. Pour, S. Clarke, M.E. Jennex, A.V. Anttiroiko, S. Kamel, I. Lee, J. Kisielnicki, A. Gupta, C.V. Slyke, J. Wang, V. Weerakkody (Eds.), Robotics: Concepts, methodologies, tools, and applications, IGI Global, 2014, pp. 1662-1720. https://doi.org/10.4018/978-1-4666-4607-0.ch081

[44] V.S.B. Kurukuru, A. Haque, M.A. Khan, S. Sahoo, A. Malik, F. Blaabjerg, A review on Artificial Intelligence applications for grid-connected solar photovoltaic systems, Energies 14 (2021) 4690. https://doi.org/10.3390/en14154690

[45] J. Feng, X. Feng, J. Chen, X. Cao, X. Zhang, L. Jiao, T. Yu, Generative adversarial networks based on collaborative learning and attention mechanism for hyperspectral image classification, Remote Sensing 12 (2020) 1149. https://doi.org/10.3390/rs12071149

[46] Y. Guo, Y. Liu, A. Oerlemans, S. Lao, S. Wu, M.S. Lew, Deep learning for visual understanding: A review, Neurocomputing 187 (2016) 27-48. https://doi.org/10.1016/j.neucom.2015.09.116

[47] M. Shehab, L. Abualigah, Q. Shambour, M.A.A. Hashem, M.K.Y. Shambour, A.I.A. Salibi, A.H. Gandomi, Machine learning in medical applications: A review of state-of-the-art methods, Computers in Biology and Medicine 145 (2022) 105458. https://doi.org/10.1016/j.compbiomed.2022.105458

[48] P. Mandal, S.T.S. Madhira, J. Meng, R.L. Pineda, Forecasting power output of solar photovoltaic system using wavelet transform and artificial intelligence techniques, Procedia Computer Science 12 (2012) 332-337. https://doi.org/10.1016/j.procs.2012.09.080

[49] V.S.B. Kurukuru, A. Haque, M.A. Khan, S. Sahoo, A. Malik, F. Blaabjerg, A review on Artificial Intelligence applications for grid-Connected solar photovoltaic systems, Energies, 14 (2021) 4690. https://doi.org/10.3390/en14154690

[50] P. Barmpoutis, P. Papaioannou, K. Dimitropoulos, N. Grammalidis, A review on early forest fire detection systems using optical remote sensing, Sensors 20 (2020) 6442. https://doi.org/10.3390/s20226442

[51] A. Mellit, G.M. Tina, S.A. Kalogirou, Fault detection and diagnosis methods for photovoltaic systems: A review, Renewable and Sustainable Energy Reviews 91 (2018) 1-17. https://doi.org/10.1016/j.rser.2018.03.062

[52] C. Ghenai, M. Bettayeb, Modelling and performance analysis of a stand-alone hybrid solar PV/Fuel Cell/Diesel Generator power system for university building, Energy 171 (2019) 180-189. https://doi.org/10.1016/j.energy.2019.01.019

[53] R. Shukla, K. Sumathy, P. Erickson, J. Gong, Recent advances in the solar water heating systems: A review, Renewable and Sustainable Energy Reviews 19 (2013) 173-190. https://doi.org/10.1016/j.rser.2012.10.048

[54] T. Adefarati, R.C. Bansal, Reliability, economic and environmental analysis of a microgrid system in the presence of renewable energy resources, Applied energy 236 (2019) 1089-1114. https://doi.org/10.1016/j.apenergy.2018.12.050

[55] B. Kroposki, Integrating high levels of variable renewable energy into electric power systems, Journal of Modern Power Systems and Clean Energy 5 (2017) 831-837. https://doi.org/10.1007/s40565-017-0339-3

[56] N. Jaloliddinova, R. Sultonov, Renewable sources of energy: Advantages and disadvantages, Достижения науки и образования 8-3 (2019) 49.

[57] M. Faisal, M.A. Hannan, P.J. Ker, A. Hussain, M.B. Mansor, F. Blaabjerg, Review of energy storage system technologies in microgrid applications: Issues and challenges, IEEE Access 6 (2018) 35143-35164. https://doi.org/10.1109/ACCESS.2018.2841407

[58] A.A.B. Rújula, Future development of the electricity systems with distributed generation, Energy 34 (2009) 377-383. https://doi.org/10.1016/j.energy.2008.12.008

[59] M.E. Meral, F. Dinçer, A review of the factors affecting operation and efficiency of photovoltaic based electricity generation systems, Renewable and Sustainable Energy Reviews 15 (2011) 2176-2184. https://doi.org/10.1016/j.rser.2011.01.010

[60] S.A. Kalogirou, Solar Energy Engineering: Processes and Systems, second ed., Academic press, 2013.

[61] A. Verma, S. Singhal, Solar PV performance parameter and recommendation for optimization of performance in large scale grid connected solar PV plant - Case study, Journal of Energy and Power Sources 2 (2015) 40-53.

[62] M.A. Hannan, M.S.H. Lipu, A. Hussain, P.J. Ker, T.M.I. Mahlia, M. Mansor, A. Ayob, M.H. Saad, Z.Y. Dong, Toward enhanced state of charge estimation of Lithium-ion batteries using optimized machine learning techniques, Scientific Reports 10 (2020) 4687. https://doi.org/10.1038/s41598-020-61464-7

[63] S. Sinha, S.S. Chandel, Review of software tools for hybrid renewable energy systems, Renewable and Sustainable Energy Reviews 32 (2014) 192-205. https://doi.org/10.1016/j.rser.2014.01.035

[64] S. Sinha, S.S. Chandel, Review of recent trends in optimization techniques for solar photovoltaic-wind based hybrid energy systems, Renewable and Sustainable Energy Reviews 50 (2015) 755-769. https://doi.org/10.1016/j.rser.2015.05.040

[65] M.A. Hannan, J.A. Ali, A. Mohamed, A. Hussain, Optimization techniques to enhance the performance of induction motor drives: A review, Renewable and Sustainable Energy Reviews 81 (2018) 1611-1626. https://doi.org/10.1016/j.rser.2017.05.240

[66] N. Sulaiman, M.A. Hannan, A. Mohamed, P.J. Ker, E.H. Majlan, W.R.W. Daud, Optimization of energy management system for fuel-cell hybrid electric vehicles: Issues and recommendations, Applied Energy 228 (2018) 2061-2079. https://doi.org/10.1016/j.apenergy.2018.07.087

[67] M.S.H. Lipu, M.A. Hannan, A. Hussain, A. Ayob, M.H.M. Saad, T.F. Karim, D.N.T. How, Data-driven state of charge estimation of lithium-ion batteries:

Algorithms, implementation factors, limitations and future trends, Journal of Cleaner Production 277 (2020) 24110. https://doi.org/10.1016/j.jclepro.2020.124110

[68] N.A.A. Razak, M.M. Othman, I. Musirin, Optimal sizing and operational strategy of hybrid renewable energy system using homer, in 4th International Power Engineering and Optimization Conference (PEOCO), 2010, pp. 495-501.

[69] T. Khatib, A. Mohamed, K. Sopian, A review of photovoltaic systems size optimization techniques, Renewable and Sustainable Energy Reviews 22 (2013) 454-465. https://doi.org/10.1016/j.rser.2013.02.023

[70] Y. Eren, İ.B. Küçükdemiral, İ. Üstoğlu, Introduction to optimization, in: O. Erdinc (Eds.), Optimization in Renewable Energy Systems, Elsevier, 2017, pp. 27-74. https://doi.org/10.1016/B978-0-08-101041-9.00002-8

[71] S.A. Kalogirou, Solar thermal collectors and applications, Progress in Energy and Combustion Science 30 (2004) 231-295. https://doi.org/10.1016/j.pecs.2004.02.001

[72] K.K. Nair, J. Jose, A. Ravindran, Analysis of temperature dependent parameters on solar cell efficiency using MATLAB, International Journal of Engineering Development and Research 4 (2016) 536-541.

[73] S. Chander, A. Purohit, A. Sharma, S.P. Nehra, M.S. Dhaka, Impact of temperature on performance of series and parallel connected mono-crystalline silicon solar cells, Energy Reports 1 (2015) 175-180. https://doi.org/10.1016/j.egyr.2015.09.001

[74] Z.A.A. Majid, M.H. Ruslan, K. Sopian, M. Othman, M. Azmi, Study on Performance of 80 Watt Floating Photovoltaic Panel, Journal Of Mechanical Engineering And Sciences 7 (2014) 1150-1156. https://doi.org/10.15282/jmes.7.2014.14.0112

[75] E. Fadhil, The Impact of the Environmental Condition on the Performance of the Photovoltaic Cell, American Journal of Energy Engineering 4 (2016) 1. https://doi.org/10.11648/j.ajee.20160401.11

[76] D.M. Tobnaghi, R. Madatov, D. Naderi, The effect of temperature on electrical parameters of solar cells, International Journal of Advanced Research in Electrical, Electronics and Instrumentation Engineering 2 (2013) 6404-6407.

[77] F. Dincer, M.E. Meral, Critical factors that affecting efficiency of solar cells, Smart Grid and Renewable Energy 1 (2010) 47-50. https://doi.org/10.4236/sgre.2010.11007

[78] R. Stropnik, U. Stritih, Increasing the efficiency of PV panel with the use of PCM, Renewable Energy 97 (2016) 671-679. https://doi.org/10.1016/j.renene.2016.06.011

[79] Y. Zheng, B.M. Jenkins, K. Kornbluth, C. Træholt, Optimization under uncertainty of a biomass-integrated renewable energy microgrid with energy storage, Renewable Energy 123 (2018) 204-217. https://doi.org/10.1016/j.renene.2018.01.120

[80] R.D. Lopez, I.R.C. Monreal, J.M. Yusta, Stochastic-heuristic methodology for the optimisation of components and control variables of PV-wind-diesel-battery stand-alone systems, Renewable Energy 99 (2016) 919-935. https://doi.org/10.1016/j.renene.2016.07.069

[81] A.M. Motaleb, S.K. Bekdache, L.A. Barrios, Optimal sizing for a hybrid power system with wind/energy storage based in stochastic environment, Renewable and Sustainable Energy Reviews 59 (2016) 1149-1158. https://doi.org/10.1016/j.rser.2015.12.267

[82] H.H. Dezaki, M. Hamzeh, H.A. Abyaneh, H.H. Khiavi, Risk management of smart grids based on managed charging of PHEVs and vehicle-to-grid strategy using Monte Carlo simulation, Energy Conversion and Management 100 (2015) 262-276. https://doi.org/10.1016/j.enconman.2015.05.015

[83] R.D. López, E.P. Cebollada, J.L.B. Agustín, I.M. Ruiz, Optimisation of energy supply at off-grid healthcare facilities using Monte Carlo simulation, Energy Conversion and Management 113 (2016) 321-330. https://doi.org/10.1016/j.enconman.2016.01.057

[84] W.A. Adheem, A. Khafory, Maximum power point tracking control of PV electronic system using neural network, in International Conference for Engineering Researches, 2017.

[85] H.A. Aribisala, Improving the efficiency of solar photovoltaic power system, University of Rhode Island, 2013.

[86] R. Sharma, S. Suhag, Novel control strategy for hybrid renewable energy-based standalone system, Turkish Journal of Electrical Engineering & Computer Sciences 25 (2017) 2261-2277. https://doi.org/10.3906/elk-1604-102

[87] G. Ding, F. Gao, S. Zhang, P.C. Loh, F. Blaabjerg, Control of hybrid AC/DC microgrids under islanding operational conditions, J. Mod. Power Syst. Clean Energy 2 (2014) 223-232. https://doi.org/10.1007/s40565-014-0065-z

[88] W.R.A. Adhem, Designing and simulating of microcontroller based on PWM Solar Charge Controller, Journal of Al-Ma'moon College, no. 19-E, 2012.

[89] M.S.H. Lipu, M.A. Hannan, A. Hussain, M.M. Hoque, P.J. Ker, M.H.M. Saad, A. Ayob, A review of state of health and remaining useful life estimation methods for

Lithium-ion battery in electric vehicles: Challenges and recommendations, J. Clean. Prod 205 (2018) 115-133. https://doi.org/10.1016/j.jclepro.2018.09.065

[90] M.A. Hannan, J.A. Ali, M.S.H. Lipu, A. Mohamed, P.J. Ker, T.M.I. Mahlia, M. Mansor, A. Hussain, K.M. Muttaqi, Z.Y. Dong, Role of optimization algorithms based fuzzy controller in achieving induction motor performance enhancement, Nat. Commun. 11 (2020) 3792. https://doi.org/10.1038/s41467-020-17623-5

[91] M.A. Hannan, Z.A. Ghani, M.M. Hoque, P.J. Ker, A. Hussain, A. Mohamed, Fuzzy logic inverter controller in photovoltaic applications: Issues and recommendations, IEEE Access 7 (2019) 24934-24955. https://doi.org/10.1109/ACCESS.2019.2899610

[92] M.K. Sahoo, P. Kale, Integration of silicon nanowires in solar cell structure for efficiency enhancement: A review, J. Materiomics 5 (2019) 34-48. https://doi.org/10.1016/j.jmat.2018.11.007

[93] S.S. Liao, Y.C. Lin, C.L. Chuang, E.Y. Chang, Efficiency enhancement of multicrystalline Silicon solar cells by inserting two-step growth thermal oxide to the surface passivation layer, International Journal of Photoenergy 2017 (2017) 9503857. https://doi.org/10.1155/2017/9503857

[94] X. Huang, S. Han, W. Huang, X. Liu, Enhancing solar cell efficiency: The search for luminescent materials as spectral converters, Chem. Soc. Rev. 42 (2013) 173-201. https://doi.org/10.1039/C2CS35288E

[95] H. Zhang, J. Toudert, Optical management for efficiency enhancement in hybrid organic-inorganic Lead halide perovskite solar cells, Sci. Technol. Adv. Mater. 19 (2018) 411-424. https://doi.org/10.1080/14686996.2018.1458578

Chapter 4
Artificial Intelligence in Material Genomics

Joy Hoskeri H[1,*], Nivedita Pujari S[1], Badrinath Kulkarni[2], Arun K. Shettar[3]

[1]Department of Bioinformatics, Karnataka State Akkamahadevi Women's University, Vijayapur-586109, Karnataka, India

[2]P.C. Jabin Science College, Vidya Nagar, Hubballi -580031, Karnataka, India

[3]Cytxon Biosolutions Pvt. Ltd., Lingaraj Nagar, Hubballi -580031, Karnataka, India

joybioinformatics@gmail.com

Abstract

The invention of new materials with desired properties is always a matter of interest. The material genome projects and material genome initiative have bought new insights into material genomics. Artificial intelligence (AI) is the decision-making ability of a computable machine. AI in material genomics has stimulated the field of material science. AI tools like Atom2vec, MATLAB, ICSD, MPIinterfaces, PyCDT, and AFLOWLIB have opened ways for the discovery of various materials. These AI tools and databases are also efficient in the property prediction of new materials, improvement in characterization protocols, experimental parameter standardization, fastening simulation scale, development of high throughput methods, and data analysis. The current chapter is focused on AI-based developments in the material genomics.

Keywords

Material Genomics, Artificial Intelligence, Machine Learning, AI Tools, and Databases

Contents

Artificial Intelligence in Material Genomics 87

1. Introduction 88
2. Material genomics 89
3. Strength of artificial intelligence 92

4. Artificial intelligence in material genomics 94

Conclusion 99

References 99

1. **Introduction**

Materials in general are an important part of science and technology. For a material to be desirable for use in different fields, it has to possess certain unique attributes such as durability, economic friendly, sustainability, etc. [1]. These properties are important for the commercialization of materials. Material science is the domain of material research that mainly deals with the design, manufacturing, innovation, and development of novel materials with a wide range of applications. Material genomics approach is an important contribution to the material genome initiative (MGI) introduced by Barak Obama, the former President of the United States [2]. The initiative mainly focused on bringing scientists and researchers involved in various disciplines together for the design and discovery of new materials. It mainly aimed at making commercialization easier for such novel materials, thus bringing the market place closer to research. Material genomics is the combination of computational tools, experimental tools, and data analysis. It makes use of virtual screening and could be used for designing and standardization *in-silico* [3]. Artificial intelligence is an emerging field that has applications in various fields. AI because of its ability to mimic human intelligence, has made several processes easier, time efficient, and also cost-efficient [4]. AI has several applications including health care, agriculture, industries, and material science [5, 6]. Two important domains, of material genomics, which include deep learning and machine learning have made numerous changes in the field of material science [7]. With the help of ML and DL, several tools, databases, and high throughput methods to solve material science problems are being discovered. These developments have made a remarkable change in material genomics. Artificial intelligence material genomics has developed tools for framing the periodic table of elements [8], predicting material properties with the aid of tools based on machine learning [9], pattern reorganization [10], and parameters optimization of materials [11]. The flow chart (Fig. 1) summarizes the functioning of material genomics with the help of artificial intelligence.

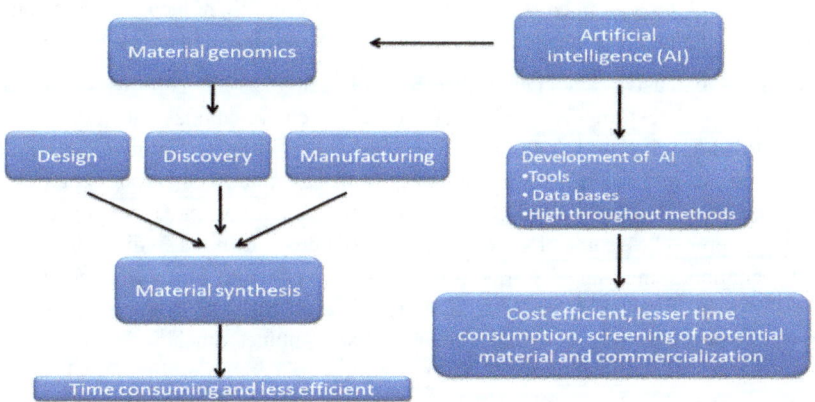

Figure 1. Flow chart depicting the contribution of AI in the acceleration of developments in materials genomics.

The current chapter is focused on development in material genomics caused due to material genome project and material genome imitative, artificial intelligence and its role in improving material genomics, various tools, datasets, and methods reported so far that help in the efficient screening of novel and potential materials.

2. Material genomics

Materials are strong grounds for technology and innovation, discovery and commercialization of new, novel, and advanced materials which play a vital role in solving various problems associated with economic growth, innovation, and the environment [1]. However, the discovery of new material includes the hit and trial method and also consumes time to standardize a material for technological application and commercialization. Therefore, an organized and scalable approach that supports both researchers and the material community is required. The concept of material genomics started in 2011 and was announced as a material genome initiative (MGI) [2]. MGI confronted the societal science and engineering platforms to speed up innovations of novel materials which help in innovative designing, manufacturing, eco-friendly machines, and many more [12]. MGI encouraged scientists to integrate it with other fields to develop genuine products at an increased rate. Material genomes were brought up and studied to innovate, research and build novel materials that can address problems and help in improving social well-being [13]. Material genome initiative or MGI was initiated by US president Barak Obama to enhance global competitiveness by reducing the time and cost to half for bringing novel materials from research labs to market place in the nation [2].

MGI is an extensive association formed with scientists of both experimental and computational backgrounds to develop scientific computational methods to screen, interpret and standardize at an unbiased scale and rate. The material genomics approach (Fig.2) consists of three important requirements such as experimental tools, computational tools, and digital data (Figure 2). Several research groups have employed material genomics approaches, some of which are discussed here: one of the studies done by Khaira et al. combined small-angle X-ray scattering, physics-based molecular modeling, and progressive standardization to frame films for experiments [14]. Another study reported the use of high-throughput virtual screening that combines machine learning, quantum chemistry, and experimental optimization to explore around 1.6 million OLED molecules [15]. One of the studies applied a combination of quantum mechanical stimulations to develop *in-silico* polar metal showing variable stability. This theory-guided material genomics approach revealed an infamous class of materials that can be important concerning innovations and technologies [16].

Research of high-throughput materials with string physiochemical properties is quite challenging, however, the material genomics approach makes use of lesser mechanical sensitivity and high-energy materials for the innovation of materials with higher energy density. Such an approach was used in a study to accelerate the process of innovation of novel insensitive high-energy explosives by distinguishing rapid genetic features, screening, and molecular design, through which they synthesized two target molecules 2,4,6-triamino-5-nitropyrimidine-1,3-dioxide [17].

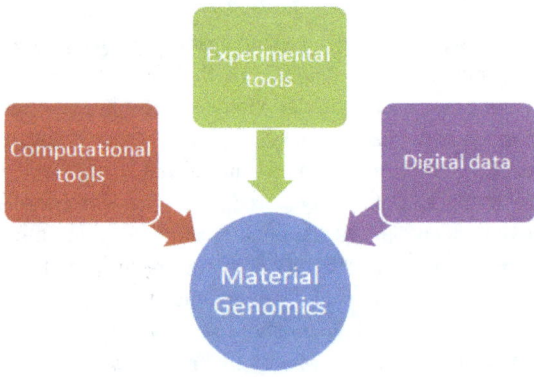

Figure 2. Material genomics approach

The material genomics approach utilizes image screening of materials that can be used for standardizing materials *in-silico* [3]. Many studies have already used this highly efficient approach to analyze thousands of compounds as efficient technological materials. Some of the putative examples are listed below in Table 1.

Table 1. Different approaches in material genomics.

Sl.No.	Examples	Reference
1	Thermoelectrics	[3]
2	Hydrogen storage material	[17]
3	Topological insulators	[1]
4	Solar photovoltaics	[18]
5	Li-ion batteries	[16]

Another recent development in material genome includes the integration of computational tools and information technology, this methodology has allowed access to different computed material datasets to various research communities and has allowed new collaborations for material discovery [19]. The material project, an important part of the material genome initiative aims to speed up material design by generating open collaborative systems that target each step in the design process (data acquisition, validation, analysis, and design) [20]. Polymer nanocomposites are a diverse class of materials, whose nano-range particles, chemistry, and polymer resin combination have made them potential compounds [21]. With the help of the material genome initiative, a data-guided web-based platform called nano-mine has been developed. Nano-mine can analyze and construct polymer nanocomposite systems under the material genome initiative concept [22].

Machine learning is a fast-growing field that has shown its way in several fields. A detailed review [23] is available on materials databases, analytical tools, and material data used in machine learning algorithms. Such an approach is designed to achieve extraordinary results in material discovery and design.

Material genomics, therefore, aims at connecting researchers of different communities with various technologies that help in bringing novel materials to the marketplace. Therefore material genomics approach can be summarized as a combination of innovation, discovery, development, and collaboration. Material genomics makes use of researchers that work on computational tools and then connect them with experimental approaches for further validation. A combined output of such an approach is analyzed using databases. Since the approach is interdisciplinary and benefits people of different backgrounds, material genomics has emerged as a potential field. An interrelationship of the material genomics approach is presented in Figure 3.

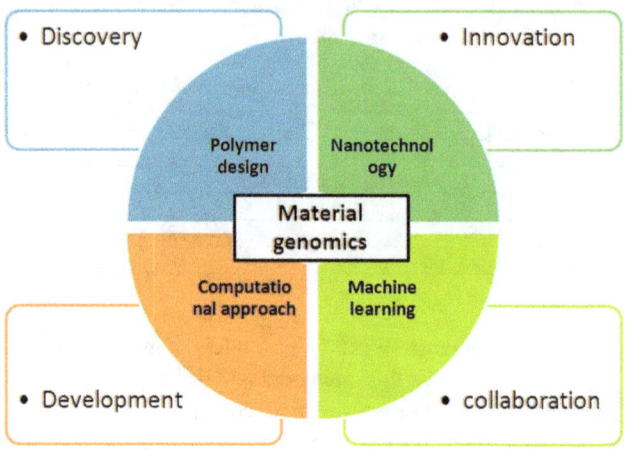

Figure 3. Interrelationship of material genomics with various fields and domains.

3. Strength of artificial intelligence

Artificial intelligence (AI) is a newly emerging branch of science and technology that is based on computer programs to perform various tasks of human needs. The field is based on the concept of – "Can a machine think?" [4]. AI refers to mimicking human intelligence by a computational system or a machine. Such decision-making technology is used in various fields including health science, machine learning, material science, etc. The first breakthrough of AI was back in 1996 when Geoffrey Hinton and his co-workers proposed an approach to building deeper neural networks [24]. This approach redefined AI research and also different algorithms like deep learning (DL) and machine learning (ML). DL is a part of ML that forms multiple layers of neural networks which represent learning [7]. However, ML is part of AI which a computer system or program can use to learn and acquire intelligence without human intervention. A few implications of artificial intelligence are given below in Table 2. (Fig.4)

Table 2. Implications of artificial intelligence.

SL. No	Area/field	Purpose	Reference
1	Intensive care unit	AI tools function as intelligent assistants to clinicians for constant monitoring and thus help in reducing costs and improving the efficiency of machines.	[5]
2	Higher education	AI tools in education are used for assessment, evaluation, profiling, and intelligent tutoring and help in the dynamic development of the education system.	[25]
3	Covid-19	AI tools and databases are used to address challenges posed by covid19 at molecular, clinical, and diagnostic levels.	[6]
4	Urology	AI tools are used for accurate prediction and analysis of large data charts to facilitate personalized and evidence-based care of patients.	[26]
5	Solid waste management	Software of AI and machine learning are used for waste characteristic prediction, prose output prediction, bin level detection, and process parameter prediction.	[27]
6	Cancer genomics	AI applications are used in testing cancer genetics and for variant calling in diagnostics.	[28]
7	Health care	AI application in health care helps in data acquisition, algorithmic transparency, and real-time assessment and also helps in preparing patients and physicians for modern and digitalized healthcare	[29]
8	Agriculture	The application of AI in agriculture aims at accuracy, high performance, flexibility, and cost-effectiveness of soil, disease, and weeds management.	[30]

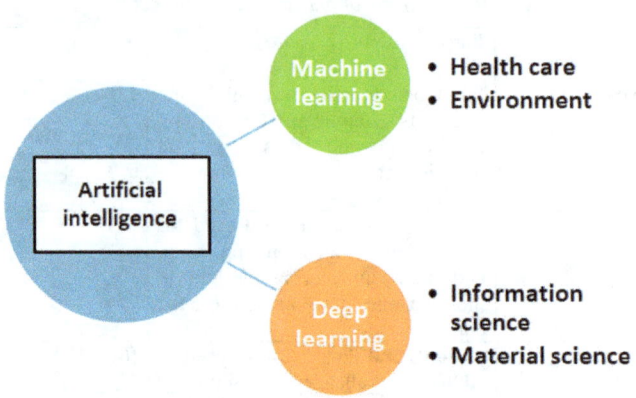

Figure 4. Domains of Artificial intelligence and their fields of applications.

4. Artificial intelligence in material genomics

The collaboration of material genomics and artificial intelligence has brought in a new dimension of studies in material science. Material genomics which mainly dealt with the design, manufacture, and discovery of new materials was brought to light by the material gnome initiative project as discussed earlier. To modernize, and initiate new approaches to the material genome, artificial intelligence was merged in. Various advances and research associated with material genomics and artificial intelligence are discussed here.

Artificial intelligence emerging with material science has brought in a new area of development in material science[31]. After a research period of over 60 years development has taken place from basic perceptron to complex neural networks[32] [33]. AI has evolved as an important algorithm framework and has proven its efficiency in various fields. Such an extraordinary data-storing ability of artificial intelligence has gained a broad range of e attention from researchers of material science. AI is a data-driven science that compresses large data information into undefined theories that guide scientific innovations [34]. Such an approach can address many problems in material research, where large-scale composite space and nonlinear processes are used. The combination of artificial intelligence with material science is defined as material informatics, it is an interdisciplinary branch that helps scientists to efficiently obtain unknown relationships between different aspects, determines specific properties of materials, optimizes different methods of chemical synthesis and process parameters and also improves the quality of current characterization methods of materials. Machine learning (ML) is an important part of AI. Material learning research in the field of material science is growing rapidly, where it has a major role in the synthesis of new materials and the prediction of various routes of synthesis [35]. The relationship between AI and material science can be explained in three generations (Figure 5). The first generation mainly deals with structure performance which is predicted by the local optimization algorithm and the performance of materials present in the structure. The second generation follows a global optimization algorithm that can carry out crystal structure elemental composition predictions. The third generation refers to statically oriented designs that make use of machine learning algorithms to find out the structure, configuration, and efficiency of chemical and physical data [36, 37]. The current research process involving the combination of material science and artificial intelligence refers to the third-generation relationship.

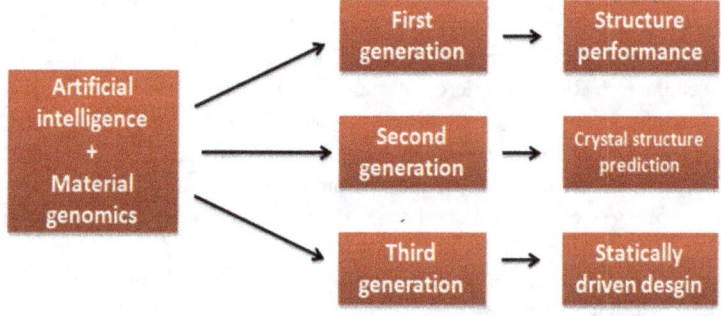

Figure 5. Generations of artificial intelligence in material genomics

Some of the important aspects of research and developments of artificial intelligence in material genomics are discussed below:

Atom2vec: is a tool for the predictive output of artificial intelligence in material science; the tool is designed on the bases of a machine learning concept, which rebuilt the periodic table in a few hours. Atom2vec works by initially differentiating atoms by investigating the series of compounds available in the online database. Later with the help of natural language processing, the characteristic of each word is derived. Chemical compounds are grouped according to the environment of the chemical. This can provide an efficient way of quantitative portrayal in the coming years [38].

Property prediction of new materials: Researchers of material science generally focus on optimizing desired properties such as electrolytes conductivity, power transformation potential of organic-inorganic hybrids, and Seebeck coefficient of thermoelectric materials [39]. The process of optimizing such properties is time-consuming and based on trial and error experiments where many times the experimental results are not satisfactory. Machine learning approach of AI devolved models that can predict shapes of materials with considerable accuracy before synthesis. One such model is MATLAB; this can be used to search for a minute amount of electrolytes that are solid among more than 12000 materials [40]. Similar methods like MATLAB are applied to design monochromatic catalysts [41, 42], light-emitting diodes [9], organic–inorganic hybrid perovskite [43], and organic light-emitting diode (OLED) [44]. Therefore, materials science has emerged not just as a trial and error method it is also efficient in reducing the number of experiments.

Improvement in characterization protocols: For material preparation using high throughput methods, material genome projects and analysis of data using AI tools have become more evident in recent years [45]. Once such development is in use of neural networks in deep learning, this approach has made considerable achievement in recognition [46]. Image characterization of micromaterials can be easily transferred by pattern recognition. Such images can be easily visualized by electron microscopy to analyze the microstructures and properties of various materials. With the help of AI, automatic defect reorganization and classification can be incorporated into electron microscopy, so that a large number of images can be screened for statically significant information. Such an approach was used by Li et al. where material information such as size and kind of defect was identified using combined machine learning, computational, and image analysis approaches [47]. Similarly, data obtained by X-ray diffraction can also be analyzed using a machine-learning approach [48].

Experimental parameter standardization: In conventional material developments, several parameters have to be considered, analyzed, and also adjusted manually during synthesis. The process is less efficient and might not find optimum parameters. The machine learning approach has the ability of nonlinear regression that can be used to search suitable areas or areas in a large diameter space [44]. One of the fields where such an approach is used is the welding process, called friction stir welding. It is a new solid-state welding process that can be used in shipbuilding, automobile, and aerospace. A study involved 108 independent experimental data for training with machine learning models that included decision trees and neural networks. The study focused on parameters of original welding such as stain rate, shear stress, temperature, and strong variables on void formation [45]. The algorithm of AI used here can find 96.6% of defects. Making use of such model standardization of parameters in the welding process can be fulfilled, and the occurrence of unfavorable parameters such as void formation in FSW from ML can be avoided. In future years, the process of material synthesis will be fully automated and combined with industries for manufacturing, for example, digital manufacturing of polymers from high-throughput programs [49].

Fastening simulation scale: machine learning approach can find out irregular repetition present in the calculation of atomic force field theoretically, and the corresponding energy can be easily calculated. Another important feature is that movement of several atoms can be enlarged into millions of atoms within a period of a few Pico-seconds. This strategy greatly increases the time range and length of simulation calculations to give better results. Apart from these complex material structures like crystallinity and amorphous, chemical reactions like interfacial reactions and corrosion can be stimulated. Such interatomic potential based on artificial neural networks has provided an unbiased way for building surface systems that are difficult to elaborate by conventional potential. A study carried out by Artrith et al. made use of zinc oxide and copper as a reference system to validate the efficiency of the interatomic strategy of artificial neural networks and explored the CuZnO ternary combination system of oxide-supported copper clusters [46].

Databases: Databases play important role in any research domain of science and technology. In the field of material science, some of the important parameters such as defects in patterns, physical properties of the compound of interest, ionization potentials, electron facilities, and numerical quantity of material properties associated with research studies are maintained in well-equipped and well-established databases. One such database is ICSD [50] database, which contains data extracted from experimental outcomes. It mainly consists of original data that are reported in various scientific articles. Later the information repositories are further screened for potential material with desired applications, with the aid of high throughput artificial intelligence methods [51]. Newly framed databases consist of sophisticated algorithms for semi-automated and automated representation and interpretation of information such as scripting language retrieval. Some of the leading databases include material project [20] and the open quantum materials database [51], which contain around 105 density functional theory (DFT) calculated crystal structures.

High throughput methods: this method of analysis is one of the prominent procedures used by researchers of several disciplines. It has been a potential tool for finding out material innovation requirements over recent years [52]. Such type of screening is commonly seen in pharmaceutical industries for screening potential compounds for drug development. However, investigation in material sciences is associated with computational tools that are employed for the synthesis of compounds experimentally, leading to the discovery of novel magnetic and dielectric materials [53]. Figure 6 explains the workflow of high throughput methods.

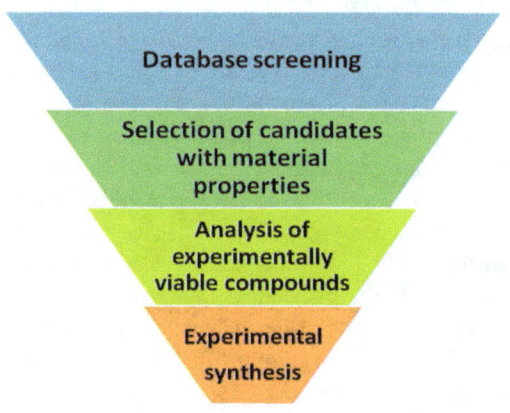

Figure 6. The flow of high-throughput material synthesis

Materials synthesized by high-throughput methods have targeted properties. AFLOWLIB is one of the huge initiatives in computed data [95]. It consists of an elaborated database of compounds, metastable structures, and binary phase diagrams computed in the AFLOWLIB workflow [54].

Data analysis: Data analysis appears as one of the limiting factors in material science in the discovery of new materials. Even though novel techniques are discovered for manufacturing and identifying numerous novel compounds and computational tools are required to analyze such large data. Various studies have proposed different software infrastructures, the majority of those developed by research groups, these support storage, collection, retrieval, and analysis of data generated by various material stimulators. Some of the examples include MPI interfaces [55]. This tool automates high-throughput computational screening and investigation of interfacial systems, which enables computational tools-based screening of nanoparticle/ligand pairs by analyzing various degrees of freedom within a system. Matminer is software that facilitates data-oriented methods for predicting and analyzing material properties [55]. PyCDT is user-friendly software and provides a direct interface with the material project database [20].

Conclusion

Material genomics is an upcoming branch of material science mainly supported by the material genome initiative, it mainly aims to bring researchers from laboratories of different material science research and market place of materials closer. However, with an increase in material science research outcomes, novel materials with varied properties and parameters are being discovered. Therefore, manual screening, analysis, and interpretation are time-consuming and also need huge financing. The intervention of artificial intelligence in combination with material learning, machine learning, and deep learning algorithms has accelerated the field of material science. AI has bought in new tools, high throughput methods, and databases for analyzing thousands of compounds to find and develop potential and functional materials and further synthesis of the same. Therefore artificial intelligence in material genomics has significantly stimulated material science research and bought in new materials of potential interest.

Declaration: All the figures and tables mentioned in the manuscript are photographed and tabulated by authors and hence no copyright permissions are required.

References

[1] K. Yang, W. Setyawan, S. Wang, M.B. Nardelli, S. Curtarolo, A search model for topological insulators with high-throughput robustness descriptors, Nat. Mater. 11 (2012) 614-619. https://doi.org/10.1038/nmat3332

[2] De Pablo, J. Juan, B. Jones, C.L. Kovacs, V. Ozolins, A.P. Ramirez, The materials genome initiative, the interplay of experiment, theory and computation, Curr. Opin. Solid State Mater. Sci. 18 (2014) 99-117. https://doi.org/10.1016/j.cossms.2014.02.003

[3] G. Hautier, A. Jain, S.P. Ong, From the computer to the laboratory: Materials discovery and design using first principles calculations, J. Mater. Sci. 47 (2012) 7317-7340. https://doi.org/10.1007/s10853-012-6424-0

[4] A.M. Turing, Computing machinery and intelligence, in: R. Epstein, G. Roberts, G. Beber (Eds.), Parsing the Turing Test, Springer, Dordrecht, 2009, pp. 23-65. https://doi.org/10.1007/978-1-4020-6710-5_3

[5] Hanson III, C. William, B.E. Marshall, Artificial intelligence applications in the intensive care unit, Crit. Care Med. 29 (2001) 427-435. https://doi.org/10.1097/00003246-200102000-00038

[6] J. Bullock, A. Luccioni, K.H. Pham, C.S.N. Lam, M.L. Oroz, Mapping the landscape of artificial intelligence applications against COVID-19, J. Artif. Intell. Res. 69 (2020) 807-845. https://doi.org/10.1613/jair.1.12162

[7] J. He, J. Chen, X. He, J. Gao, L. Li, L. Deng, M. Ostendorf, Deep reinforcement learning with a natural language action\ space, arXiv preprint arXiv 1511.04636 (2015). https://doi.org/10.18653/v1/P16-1153

[8] S. Ye, B. Li, Q. Li, H.-P. Zhao, X.-Q. Feng, Deep neural network method for predicting the mechanical properties of composites, Appl. Phys. Lett. 115 (2019) 161901. https://doi.org/10.1063/1.5124529

[9] Y. Zhuo, A.M. Tehrani, A.O. Oliynyk, A.C. Duke, J. Brgoch, Identifying an efficient, thermally robust inorganic Phosphor host via machine learning, Nat. Commun. 9 (2018) 1-10. https://doi.org/10.1038/s41467-017-02088-w

[10] Wang, Y.Tang, B. Li, X.-J. Xu, H.-B. Ren, J.-Y. Yin, H. Zhu, Y.H. Zhang, FTIR spectroscopy coupled with machine learning approaches as a rapid tool for identification and quantification of artificial sweeteners, Food Chem. 303 (2020) 125404. https://doi.org/10.1016/j.foodchem.2019.125404

[11] K.A. Severson, P.M. Attia, N. Jin, N. Perkins, B. Jiang, Z. Yang, M.H. Chen, M. Aykol, P.K. Herring, D. Fraggedakis, M.Z. Bazant, S.J. Harris, W.C. Chueh, R.D. Braatz, Data driven prediction of battery cycle life before capacity degradation, Nat. Energy. 4 (2019) 383-391. https://doi.org/10.1038/s41560-019-0356-8

[12] A. Dima, S. Bhaskarla, C. Becker, M. Brady, C. Campbell, P. Dessauw, R. Hanisch, U. Kattner, K. Kroenlein, M. Newrock, A. Peskin, R. Plante, S.Y. Li, P.F. Rigodiat, G.S. Amaral, Z. Trautt, X. Schmitt, J. Warren, S. Youssef, Informatics infrastructure for the materials genome initiative, Jom. 68 (2016) 2053-2064. https://doi.org/10.1007/s11837-016-2000-4

[13] Lu, X.-Gang, Remarks on the recent progress of Materials Genome Initiative, Sci. Bull. 60 (2015) 1966 1968. https://doi.org/10.1007/s11434-015-0937-2

[14] G. Khaira, M. Doxastakis, A. Bowen, J. Ren, H.S. Suh, T.S. Peretz, X.Chen, C. Zhou, A.F. Hannon, N.J. Ferrier, V. Vishwanath, D.F. Sunday, R. Gronheid, R.J. Kline, J.J. De Pablo, P.F. Nealey, Derivation of multiple covarying material and process parameters using physics-based modeling of X-ray data, Macromolecules. 50 (2017) 7783-7793. https://doi.org/10.1021/acs.macromol.7b00691

[15] G. Bombarelli, Rafael, J.A.-Iparraguirre, T.D. Hirzel, D.Duvenaud, D.Maclaurin, M.A. B.-Forsythe, H. S. Chae, M. Einzinger, D.G. Ha, T. Wu, G. Marcopoulos, S. Jeon, H. Kang, H. Miyazaki, M. Numata, S. Kim, W. Huang, S.I. Hong, M. Baldo, R.P. Adams, A.A. Guzik, Design of efficient molecular organic light-emitting diodes by a high-throughput virtual screening and experimental approach, Nat. Mater. 15 (2016) 1120-1127. https://doi.org/10.1038/nmat4717

[16] T.H. Kim, D. Puggioni, Y. Yuan, L. Xie, H. Zhou, N. Campbell, P.J. Ryan, Y. Choi, J.W. Kim, J.R. Patzner, S. Ryu, J.R. Podkaminer, J. Irwin, Y. Ma, C.J. Fennie, M.S. Rzchowski, X.Q. Pan, V. Gopalan, J.M. Rondinelli, C.B. Eom, Polar metals by geometric design, Nature. 533 (2016) 68-72. https://doi.org/10.1038/nature17628

[17] Y. Wang, Y. Liu, S. Song, Z. Yang, X. Qi, K. Wang, Y. Liu, Q. Zhang, Y. Tian, Accelerating the discovery of insensitive high-energy-density materials by a materials genome approach, Nat. Commun. 9 (2018) 1-11. https://doi.org/10.1038/s41467-017-02088-w

[18] L. Yu, A. Zunger, Identification of potential photovoltaic absorbers based on first-principles spectroscopic screening of materials, Phys. Rev. Lett. 108 (2012) 068701. https://doi.org/10.1103/PhysRevLett.108.068701

[19] J. Hachmann, R.O. Amaya, S.A. Evrenk, C.A. Bedolla, R.S.S. Carrera, A.G. Parker, L.Vogt, A.M. Brockway, A.A. Guzik, The Harvard clean energy project: Large-scale computational screening and design of organic photovoltaics on the world community grid, J. Phys. Chem. Lett. 2 (2011) 2241-2251. https://doi.org/10.1021/jz200866s

[20] A. Jain, S.P. Ong, G. Hautier, W.Chen, W.D. Richards, S. Dacek, S. Cholia, D. Gunter, D. Skinner, G. Ceder, K.A. Persson, Commentary: The Materials Project: A materials genome approach to accelerating materials innovation, APL Mater. 1 (2013) 011002. https://doi.org/10.1063/1.4812323

[21] Y.T. Liang, B.K. Vijayan, K.A. Gray, M.C. Hersam, Minimizing graphene defects enhances titania nanocomposite based photocatalytic reduction of CO_2 for improved solar fuel production, Nano Lett. 11 (2011) 2865-2870. https://doi.org/10.1021/nl2012906

[22] H. Zhao, X. Li, Y. Zhang, L.S. Schadler, W. Chen, L.C. Brinson, Perspective: NanoMine: A material genome approach for polymer nanocomposites analysis and design, APL Mater. 4 (2016) 053204. https://doi.org/10.1063/1.4943679

[23] Y. Liu, C. Niu, Z. Wang, Y. Gan, Y. Zhu, S. Sun, T. Shen, Machine learning in materials genome initiative: A review, J. Mater. Sci. Technol. 57 (2020) 113-122. https://doi.org/10.1016/j.jmst.2020.01.067

[24] G.E. Hinton, R.R. Salakhutdinov, Reducing the dimensionality of data with neural networks, Sci. 313 (2006) 504-507. https://doi.org/10.1126/science.1127647

[25] O.Z. Richter, V.I. Marín, M. Bond, F. Gouverneur, Systematic review of research on artificial intelligence applications in higher education-where are the educators?, Int. J.

Educ. Technol. High. Educ. 16 (2019) 1-27. https://doi.org/10.1186/s41239-019-0132-7

[26] J. Chen, D. Remulla, J.H. Nguyen, Y. Liu, P. Dasgupta, A.J. Hung, Current status of artificial intelligence applications in urology and their potential to influence clinical practice, BJU Int. 124 (2019) 567-577. https://doi.org/10.1111/bju.14852

[27] M. Abdallah, M.A. Talib, S. Feroz, Q. Nasir, H. Abdalla, B. Mahfood, Artificial intelligence applications in solid waste management: A systematic research review, J. Waste Manag. 109 (2020) 231-246. https://doi.org/10.1016/j.wasman.2020.04.057

[28] J. Xu, P. Yang, S. Xue, B. Sharma, M.S. Martin, F. Wang, K.A. Beaty, E. Dehan, B. Parikh, Translating cancer genomics into precision medicine with artificial intelligence: Applications, challenges and future perspectives, Hum. Genet. 138 (2019) 109-124. https://doi.org/10.1007/s00439-019-01970-5

[29] F. Jiang, Y. Jiang, H. Zhi, Y. Dong, H. Li, S. Ma, Y. Wang, Q. Dong, H. Shen, Y. Wang, Artificial intelligence in healthcare: Past, present and future, Stroke Vasc. Neurol. 2 (2017) 000101. https://doi.org/10.1136/svn-2017-000101

[30] E. Chukwu, N. Clara, Applications of artificial intelligence in agriculture: A review, Eng. Technol. Appl. Sci. Res. 9 (2019) 4377-4383. https://doi.org/10.48084/etasr.2756

[31] R. Kline, Cybernetics, automata studies and the Dartmouth conference on artificial intelligence, IEEE Ann. Hist. Comput. 33 (2010) 5-16. https://doi.org/10.1109/MAHC.2010.44

[32] McCulloch, S. Warren, W. Pitts, A logical calculus of the ideas immanent in nervous activity, Bull. Math. Biophys. 5 (1943) 115-133. https://doi.org/10.1007/BF02478259

[33] D.T. Tran, S. Kiranyaz, M. Gabbouj, A. Iosifidis, Heterogeneous multilayer generalized operational perceptron, IEEE Trans. Neural Netw. Learn. Syst. 31 (2019) 710-724. https://doi.org/10.1109/TNNLS.2019.2914082

[34] D. Silver, A. Huang, C.J. Maddison, A. Guez, L. Sifre, G.V.D. Driessche, J. Schrittwieser, et al., Mastering the game of Go with deep neural networks and tree search, Nature. 529 (2016) 484-489. https://doi.org/10.1038/nature16961

[35] J.J. Möller, W. Körner, G. Krugel, D.F. Urban, C. Elsässer, Compositional optimization of hard magnetic phases with machine-learning models, Acta Mater. 153 (2018) 53-61. https://doi.org/10.1016/j.actamat.2018.03.051

[36] M. Schmidt, H. Lipson, Distilling free-form natural laws from experimental data, Science. 324 (2009) 81-85. https://doi.org/10.1126/science.1165893

[37] H. Salmenjoki, M.J. Alava, L. Laurson, Machine learning\ plastic deformation of crystals, Nat. Commun. 9 (2018) 1 7. https://doi.org/10.1038/s41467-018-07737-2

[38] de Almeida, A. Filipa, R. Moreira, T. Rodrigues, Synthetic organic chemistry driven by artificial intelligence, Nat. Rev. Chem. 3 (2019) 589-604. https://doi.org/10.1038/s41570-019-0124-0

[39] N. Schneider, D.M. Lowe, R.A. Sayle, G.A. Landrum, Development of a novel fingerprint for chemical reactions and its application to large-scale reaction classification and similarity, J. Chem. Inf. Model. 55 (2015) 39-53. https://doi.org/10.1021/ci5006614

[40] A.D. Sendek, E.D. Cubuk, E.R. Antoniuk, G.Cheon, Y.Cui, E.J. Reed, Machine learning-assisted discovery of solid Li ion conducting materials, Chem. Mater. 31 (2018) 342-352. https://doi.org/10.1021/acs.chemmater.8b03272

[41] E. Kim, K. Huang, A. Saunders, A. McCallum, G. Ceder, E. Olivetti, Materials synthesis insights from scientific literature via text extraction and machine learning, Chem.Mater. 29 (2017) 9436-9444. https://doi.org/10.1021/acs.chemmater.7b03500

[42] M. Sun, T. Wu, Y. Xue, A.W. Dougherty, B. Huang, Y. Li, C.H. Yan, Mapping of atomic catalyst on graphdiyne, Nano Energy 62 (2019) 754-763. https://doi.org/10.1016/j.nanoen.2019.06.008

[43] S. Lu, Q. Zhou, Y. Ouyang, Y. Guo, Q. Li, J. Wang, Accelerated discovery of stable Lead-free hybrid organic inorganic perovskites via machine learning, Nat. Commun. 9 (2018) 1-8. https://doi.org/10.1038/s41467-017-02088-w

[44] P.M. Attia, A. Grover, N. Jin, K.A. Severson, T.M. Markov, Y.H. Liao, M.H. Chen, B. Cheong, N. Perkins, Z. Yang, P.K. Herring, M. Aykol, S.J. Harris, R.D. Braatz, S. Ermon, W.C. Chueh, Closed loop optimization of fast-charging protocols for batteries with machine learning, Nature 578 (2020) 397-402. https://doi.org/10.1038/s41586-020-1994-5

[45] Y. Du, T. Mukherjee, T. DebRoy, Conditions for void formation in friction stir welding from machine learning, NPJ Comput. Mater. 5 (2019) 1-8. https://doi.org/10.1038/s41524-019-0207-y

[46] A. Krizhevsky, I. Sutskever, G.E. Hinton, ImageNet classification with deep convolutional neural networks, Adv. Neural Inf. Process. Syst. 25 (2012) 84-90. https://doi.org/10.1145/3065386

[47] B. Lin, J.L. Hedrick, N.H. Park, R.M. Waymouth, Programmable high-throughput platform for the rapid and scalable synthesis of polyester and polycarbonate libraries, J. Am. Chem. Soc. 141 (2019) 8921-8927. https://doi.org/10.1021/jacs.9b02450

[48] B. Lin, J.L. Hedrick, N.H. Park, R.M. Waymouth, Programmable high-throughput platform for the rapid and scalable synthesis of polyester and polycarbonate libraries, J. Am. Chem. Soc. 141 (2019) 8921-8927. https://doi.org/10.1021/jacs.9b02450

[49] M. Ziatdinov, A. Maksov, S.V. Kalinin, Learning surface molecular structures via machine vision, NPJ Comput. Mater. 3 (2017) 1-9. https://doi.org/10.1038/s41524-017-0038-7

[50] A. Belsky, M. Hellenbrandt, V.L. Karen, P. Luksch, New developments in the Inorganic Crystal Structure Database (ICSD): Accessibility in support of materials research and design, Acta Crystallogr: B, Struct. Sci. 58 (2002) 364-369. https://doi.org/10.1107/S0108768102006948

[51] J.E. Saal, S. Kirklin, M. Aykol, B. Meredig, C. Wolverton, Materials design and discovery with high throughput density functional theory: The open quantum materials database (OQMD), Jom. 65 (2013) 1501-1509. https://doi.org/10.1007/s11837-013-0755-4

[52] N. Mounet, M. Gibertini, P. Schwaller, D. Campi, A. Merkys, A. Marrazzo, T. Sohier, I.E Castelli, A. Cepellotti, G. Pizzi, N. Marzari, Two dimensional materials from high-throughput computational exfoliation of experimentally known compounds, Nat. Nanotechnol. 13 (2018) 246-252. https://doi.org/10.1038/s41565-017-0035-5

[53] I. Takeuchi, R.B. Dover, H. Koinuma, Combinatorial synthesis and evaluation of functional inorganic materials using thin film techniques, MRS Bull. 27 (2002) 301-308. https://doi.org/10.1557/mrs2002.97

[54] S. Curtarolo, W. Setyawan, G.L.W. Hart, M. Jahnatek, R.V. Chepulskii, R.H. Taylor, S. Wang, J. Xue, K. Yang, O. Levy, M.J. Mehl, H.T. Stokes, D.O. Demchenko, D. Morgan, AFLOW: An automatic framework for high-throughput materials discovery, Comput. Mater. Sci. 58 (2012) 218-226. https://doi.org/10.1016/j.commatsci.2012.02.005

[55] M. Morales, J. Paul, A. Fumarola, V. Taliaronak, A. Shirsekar, J. Kidner, Z. Ali, M. Ali, Electronic band structure screening for Dirac points in Heuslers, arXiv preprint arXiv:2205.02547 (2022).

Chapter 5

Applications of Artificial Intelligence in Polymer Manufacturing

Satyansh Srivastava[1], Bhoomika Varshney[1], V.P. Sharma[2,*] Babra Ali[3]

[1]Kalinga Institute of Industrial Technology, Bhubaneswar, Odisha-751024, India

[2]CSIR-Indian Institute of Toxicology Research, Lucknow- 226001, UP, India

[3]Birla Institute of Technology And Science–Pilani (BITS–Pilani)(WILP), Vidya Vihar, Pilani, Rajasthan 333031

*vpsitrc1@rediffmail.com

Abstract

Artificial Intelligence (AI) is creating an everlasting impact on science and benefiting in realizing new and revolutionary sustainable materials. The field of manufacturing polymers has witnessed more exponential growth than ever, and the credit entirely goes to the novel artificially intelligent machine learning strategies. Artificial Intelligence is progressing in the interdisciplinary field of improving the lifestyle and attaining sustainable development goals (SDGs). It may be transforming applications ranging from new materials to personalized medicine and precise sensor developments. Polymer informatics is an interdisciplinary field of research converging polymer science with computer science, information science, and machine learning, serving as a muse to transform the field of polymer manufacturing altogether. With the tremendous upsurge of data in science, data-driven strategies are being utilized in polymer informatics for better and more efficient development, design, and discovery of polymers. We have attempted to discuss the application of artificial intelligence in different sectors such as polymeric designing, the food industry, healthcare, cosmetics, and agricultural sustainable productivity.

Keywords

Intelligence, Innovation, Machine Learning, Polymer, Sustainable

Contents

Applications of Artificial Intelligence in Polymer Manufacturing 105

1. Introduction 106
   1.1 Advantages and disadvantages of artificial intelligence in
       polymer manufacturing 108
2. Classification of artificial intelligence 108
   2.1 Classification of AI based on capabilities 109
3. Key Developments and commercialization in the polymer industry 110
4. Application of artificial intelligence in polymer manufacturing 111
   4.1 Artificial intelligence and polymer manufacturing 111
   4.2 Biodegradable polymers and artificial intelligence 112
   4.3 Artificial intelligence and packaging industries 113
   4.4 Agriculture and artificial intelligence 113
   4.5 Healthcare and artificial intelligence 114
   4.6 Artificial intelligence and dentistry 114
   4.7 Food industry and artificial intelligence 115
   4.8 Cosmetic artificial intelligence 115
5. Future prospects and conventional challenges 116
6. Guidelines, rules, and regulations for polymeric manufacturing 117
Conclusion 117
Acknowledgment 118
Conflict of Interest 118
Reference 118

1. Introduction

A new application of artificial intelligence (AI) and machine learning (ML) may help to achieve global sustainability goals by exponentially increasing world connectivity. AI also needs to mitigate the risk and challenges of cyber security via technology and education [1]. In the words of Bill Gates, the power of AI is incredible, it will change society in some very deep ways. Artificial intelligence is a fast-developing interdisciplinary science with great popularity and promises to originate from significant advances through business-driven hype and partially unrealistic expectation. Machine learning uses algorithms for the derivation of feature patterns from training data to classify the test objects to address regression tasks. This is generally statistical and assists to derive predictive models based on inference from linear-nonlinear instances which may help to achieve global sustainability goals through exponentially increasing world connectivity.

AI is a field of computer science that combines the knowledge of computer science, machine, and deep learning, and robust databases to solve any problem. It is based on computers and machines that mimic the human mind. After the use of AI, computers, and machines are capable of taking decisions and buildup problems solving capacity. This is an algorithm-based approach [2] and tries to imitate what actual intelligence might produce by learning from data or looking for patterns from some unlabeled data source to drive expressive insights from the same, without including any meaningful bias or emotions into the produced results. It is considered as a capability of systems to interpret the data obtained from external sources and utilized through learning for attaining specific tasks viz flexible adaptations. AI is used in almost every domain of science and technology to create an impact on polymeric materials. It is impacting the simulation of human intelligence processes through machines or computer-based systems. It includes expert systems, language processing, reorganization of speech or vision, and mindset. Machine learning is a widely and most developing field of artificial intelligence. It uses probabilistic, statistical algorithm methods to detect the experimental result. The outcome of machine learning is based on past experiences [3]. The rapid development of artificial intelligence such as the Internet of Things (IoT), 4G, 5G, machine learning, and 9D technologies has stimulated the development of digital, intelligence and the latest manufacturing technologies such as micro-bioreactor. Several tools and algorithms are used in bioinformatics such as the basic local alignment search tool(BLAST), [4]. The information industry may also generate a new delivery platform for Machine Learning sets to considerably accelerate the adoption delivery accelerate adaption of new revenue streams [5].

Polymeric compounds cover a wide spectrum of applications in different sectors such as energy storage, construction, biomedical, aerospace, 3D printing, and so on. The structural complexity of bio-polymeric compounds has become a challenge to their development, evaluation, and access [6]. The structural design and properties of a polymeric compound have been enhanced by Artificial Intelligence and their accessibility is also not limited to a place. It opens a new way for polymeric science and engineering and discovers polymeric compounds not only in the old databases but also helps the development of new polymeric compounds with the use of old databases [7]. In the modern scenario, AI is widely used during the manufacturing of polymers and their derivates. Polymers are natural or synthetically manufactured substances that hold a plethora of utilization in modern society. During polymer formation, simpler units called a monomer of one or different types form rather complex chains [8]. Monomer units contribute to the formation of highly diverse materials, both physically and chemically. The intriguing characteristics of polymers have driven an extensive amount of research in the field of material science to understand and design novel polymeric structures that can cater to the requirements of high-functioning materials [9].

Artificial Intelligence (AI) is a future technology expected to develop under an advanced network infrastructure with the integration of control technology, virtual and real artificial intelligence, and machine learning. Machine learning is the computational technique that is used to improve and enhance monitoring, maintenance, production scheduling, and the product's manufacturing. Machine learning has probabilistic prospects and utilizes the tools of probability theory which has been the strength of statistics, data mining, and engineering. AI plays a very important role in moving traditional industries towards the industrial revolution, e.g., Industries 4.0. It is the 4th industrial revolution and we must understand distribution patterns given market demands with the help of cloud computing and internet facilities [10].

1.1 Advantages and disadvantages of artificial intelligence in polymer manufacturing

There are several advantages and disadvantages for the industries to move towards AI. Some are given below [48, 49, 50].

S.No	Advantages	Disadvantages
1	Advanced and Intelligent analytical features	High Cost
2	High working capacity	Fear of job replacement
3	Fast decision-making capacity	No clear strategy for manufacturing
4	24*7 availability	Unemployment
5	High optimizing capacity	Fewer chances for creativity

2. Classification of artificial intelligence

Artificial Intelligence can be divided into different categories depending on its capabilities, functionality, and so on. Here we discuss the types of artificial intelligence according to their functionality and capabilities, Figure 1 illustrates the types of Artificial Intelligence [11].

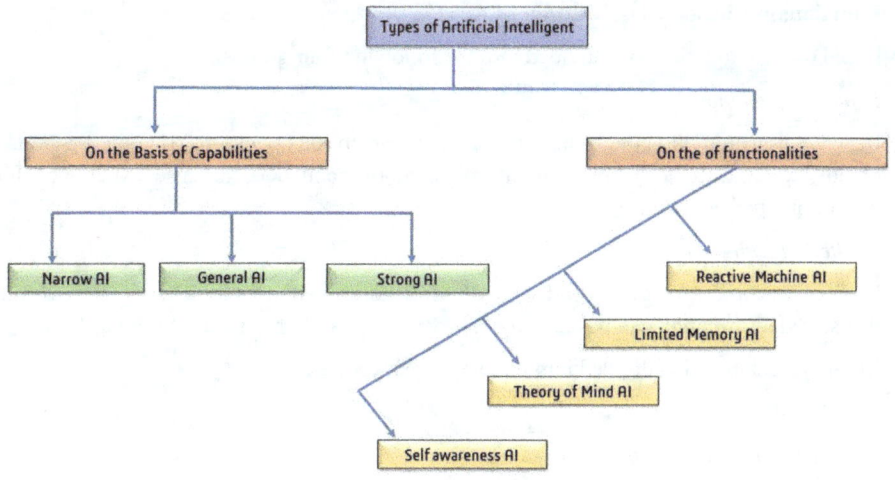

Figure 1: Classification of artificial intelligence

2.1 Classification of AI based on capabilities

Based on their capabilities, AI is classified into three subcategories.

Narrow AI

Narrow AI is also known as Artificial Narrow Intelligence (ANI). It is capable of performing one narrow task with a set algorithm and it is not so intelligent to perform self-tasks. The most common and currently available artificial intelligence is narrow AI. It is a building block for general and strong artificial intelligence.

General AI

General AI is also known as Artificial General Intelligence (AGN). It is a developing field of AI. In this field, machines will be capable of thinking and making decisions just like humans. Presently, this technology is not in existence.

Super AI

Super AI also known as Artificial Super Intelligence (ASI), is the most advanced form of artificial intelligence. This super AI machine will suppress humans. Presently, this AI is hypothetical intelligence.

Functionality-Based Classification of AI

Based on functionality AI is divided into the following four groups.

Reactive Machine AI

Reactive Machine AI is the basic form and performs basic operations. It is the first stage of any AI task. It does not contain any memories and does not use experience for performing present tasks.

Limited Machine AI

Limited Machine AI is the second stage of AI. It stores memories for a certain period, and this stored information can be used to predict the results of present and future tasks.

Limited machine AI is divided into three major categories.

- Reinforcement Learning
- Long short-term Memory
- Evolutionary Generative Adversarial Network

Theory of Mind AI

This is the 3rd stage of AI. Researchers have reached the beginning phase of the theory of mind AI. This is an advanced form of AI. In this type of AI, machines are capable of understanding emotions, beliefs, and expectations and can interact with society.

Self-awareness AI

This is the most advanced form of AI. In this AI, machines have their consciousness, are super intelligent, and think like humans.[12,13].

Types of Machine Learning

It may be classified into the following categories [15,16]

- Predictive or supervised learning
- Descriptive or unsupervised learning
- Reinforcement learning

3. Key Developments and commercialization in the polymer industry

The significant developments of artificial intelligence provide an opportunity for the production of new-generation polymeric materials with sustainability and the use of innovative polymerization mechanisms.

There are several methods such as computer-controlled flow reactors, online benchtop, NMR, and gel permeation chromatography (GPC) which are used to obtain target polymeric compounds. Reversible addition–fragmentation chain transfer (RAFT) technology works on the Thompson sampling efficient multi-objective optimization (TSEMO) algorithm. This algorithm is based on the trade-off between molar-mass dispersity and mono conversion without the interference of the user.

Novel scientific methods are now replacing the outdated empirical and experiment-based methods to design polymer molecules on both the macro and micro scales to meet the ever more demanding needs of the advancing technologies. Despite the remarkable growth in the development of the polymer industry, continual improvement is needed in the chemical properties of polymers [17, 18].

Energy: Polymers are essential for shifting to alternative energy generators or fuel cells with high durability and efficiency [19].

Sustainability: The green and cleaner substitute of polymeric composites provides techno-feasible competitiveness with reduced damage to the environment. The twelve principles of green technology are adopted for attaining sustainability [20]

Health Care: Artificial heart valves, contact lenses, dental composites, and a slew of other medical devices would have been difficult to develop if polymers had not been available. Despite these advancements, the full impact of polymers has yet to be explored. Polymer science's advancements in drug development and delivery, nano-medicine, genomics, cell differentiation, and behavior control show that polymers have much more to offer in the healthcare industry[21].

4. Application of artificial intelligence in polymer manufacturing

4.1 Artificial intelligence and polymer manufacturing

The design of polymers and manufacturing, of composites, are exploring advanced analytical tools using artificial intelligence and machine technologies. It may help in the digitalization of material design and processes of manufacturing smart materials for industry 4.0 [21,22]. The substantial time and cost in the development of a new polymer through the identification of promising molecules with pre-known physiochemical properties help to meet arbitrary given requirements [24]. The systemic integration of material design, synthesis, characterization, etc may revolutionize. They have potential applications in aerospace, mechanical, biomedical, civil engineering, and textile design. Moreover, the advanced manufacturing methods viz. stereo lithography [25], direct inciting, fused disposition modeling, laser sintering in polymer blending, etc. have raised the expectation of aerospace 3D printing from $ 1.38 billion in 2020 to $ 6.80 by 2030 with the registration of compounded annual growth rate CAGR of 18%. The aerospace market benefits from printing a complex component in one piece with the possibility of replacing several small joints and parts which leads to design optimization and rapid prototyping [26].

Polymeric materials may be designed with specific characteristics to stimulate viz. light, electricity, heat, pH [27] which have attractions in the energy environment and biomedical sectors [28]. Thus, polymers are the most exploited material for the construction of nanobiohybrides [29] and the manipulation of the biological properties of living organisms. Nanobiohybrides endow new or augmented properties which may be innate or exogenous such as stress tolerance, metabolisms programming, proliferation, conductivity, or artificial biochemical mechanisms. Because of this advancement in materials, processing technologies are making it feasible to tailor the physiochemical properties of new materials coupled with biological systems [30]. The polymeric carriers may be designed as non-viral vector and tailored for required functionalities. In these directions for efficient gene delivery [31], novel amphiphilic co-polymers to form stable micelles and Inter-polyelectrolyte complexes are in the formative stages. The interdisciplinary high-tech applications include energy harvesting, biocatalysts, bio-sensing, medicine, robotics, etc. The key developments in the plastic and polymer industry are catering to the increasing demands in defense, manufacturing, logistics, and other sectors.

4.2 Biodegradable polymers and artificial intelligence

Biodegradable polymers are critical attractions in polymer informatics, and how they can help in innumerable ways to aid different industries like food packaging [32], bio-implants, textiles, and so on. Regulatory policies are structured to inhibit the exploitation of the new field and its future application. The researchers are developing a machine learning method for the creation of a new variance of an enzyme (e.g. Fast – PETase) which may be helpful for the degradation of plastics in a few hours or days. This would be helpful for the cleaning of landfills.

It is expected that the functional, active, tolerant polyethylene terephthalates (PETase) to depolymerize e.g., breakdown into small parts at low temperatures in this direction scientists have cultured *Ideonella sakaiensis* to degrade in approx. 6 weeks at 30°C ±2 [33]. The application of bioactive hydrogel dressing with prolonged antimicrobial activity [34] may be extremely helpful in cases of burn wound healing to prevent scarring for the development of injectable [35] hydrogelspolyvinyl alcohol and microspheres containing vancomycin, gentamicin [36] or other healing agents may be used by experts of the medical fields. The experimentation and stimulation of advanced composites and hybrid materials are exponentially increasing the integration of discoveries from varied disciplines to investigate structure property performance relationships.

4.3 Artificial intelligence and packaging industries

The Europeans union directive 2019/ 904 emphasizes the reduction of the impact of certain plastic products on the environment and human health. The global issue of the overproduction of plastics with a negative impact on consumer health can be taken up through artificial intelligence tools. The specific categories of plastics may be restricted to minimization of adverse implications on bio-systems. We need to understand the trends and biodegradability patterns of disposable food packaging materials. The active packaging system may help to reduce the wastage of food by providing internal barriers to external conditions and improvement in functionalities associated with food preservation, scavenging, releasing, imitating properties, and quality control. This is feasible through a better understanding of interactions between packaging, environment, and food for attaining, commercial translation [37, 38].

4.4 Agriculture and artificial intelligence

In the domain of agriculture, the development of wireless sensor networks has improved the design of critical devices with the Internet of Things (IoT). This is an acceptable tool for automating and making potential decisions. We may integrate the sensor networks with AI systems for the assessment of land suitability for agricultural purposes. Multilayer perceptions for the assessment of agriculture land suitability have been extensible. Machine learning is one of the cutting technologies for precision agriculture to address recent challenges of agricultural suitability. The soil parameters viz organic carbon and moisture detection of diseases and weeds in the crop may be easily done through machine learning. Moreover, these approaches may be integrated for livestock assessment, predictive fertility patterns, cattle behaviors, etc. The intelligent irrigation and harvesting concepts may also be demonstrated towards sustainable productivity and quality assessment of the agricultural outputs [39].

The utility of AI in several areas of interdisciplinary science has multiplied exponentially with deep learning concepts for the integration of non-invasive techniques for robust analysis. The plant phenomics approaches which relate to plant growth, tolerance, and ecological variances e.g., complex plant rates are significant indicators. Artificial intelligence may be used for quality assurance a safety of agricultural produce with great efficiency and accuracy [40].

4.5 Healthcare and artificial intelligence

Artificial intelligence and IoT may help to completely change the polymeric business market share in the field of biomedical devices through the advancement of indigenous technologies for devices and materials used for cardiac, dentistry, and neurological disorders. In India, steps are being taken for developing innovative polymeric products for the especially challenging personnel for improving their quality of life.

Cartilage tissue engineering is an expensive and time-consuming affair with the requirement of numerous trials. Medical knowledge is improving through artificial intelligence and machine learning-assisted cardiac diagnosis /prognosis has been a boon for patients with cardiovascular diseases. Globally heart diseases and stocks are common causes for more than 80% of cardiovascular diseases(CVDs) moralities. The high level of biomarkers in the sample of human blood may be associated with pain in the chest and variances in electrocardiogram (ECG) may be supplemented for effective management. Artificial Intelligence can be easily identified as subtle patterns from data to use for medical purposes. Software programmers are helpful to diagnose medical problems through real-time processes or datasets. Few researchers have created a system made from networks of tinning polymeric fibers to replicate the in-site function of the human body to serve as the organic transistor. These networks may be helpful to detect and classify abnormal electrical signals in the body, irregular heartbeats, and significant variances in blood sugar. The network of polymer fibers may be made up of carbon-based material e.g., poly3,4 (ethylene dioxythiofine) polystyrene sulphonate (PEDOT) [41], which is a conducting polymer and used to develop soft actuators or artificial muscles and look like neurons. However, these materials are still in the earlier stages of their development with limitations of complete understanding. The AI-based software may explore unprecedented avenues for speeding up drug discovery and providing effective drugs to patients with high precision.

4.6 Artificial intelligence and dentistry

Artificial Intelligence in bio-implant planning facilities the work of the medical fraternity to support mechanism implantology. The deep approach has been successfully used for dental implants through cone-computed tomography images. The polymeric resins, materials and films are finding increased applications for restoring and replacement of tooth structure and missing teeth. Several dental materials (composites, film impression materials, denture-based reins restorative) and tissue regeneration procedures have been preferred due to low production cost, ease of processing, and easy molding for a wide range of applications. The rapid advances in technology and computed-assisted cognitive functions may help in arthroplasty. Machine learning may mimic the neuronal connection of the brain by creating an artificial neuron network (ANN). The recent advancements include computing power, learning algorithms, data storage, and the feasibility of data sources from electronic medical records and health trackers. Thus, artificial intelligence is extensively used across several fields of respiratory, gastroenterology, oncology, nephrology, ophthalmology and cardiology, etc [42].

4.7 Food industry and artificial intelligence

Artificial intelligence plays a vital role in the prediction of the parameters for quality and quantity control of food. Machine learning, neural network, and adaptive neuro-fuzzy inference system (ANFIS) are some algorithms being used in food and allied industries to fulfill consumer demand. The irregular availability of raw materials and feedstock management is one of the critical challenges for food processors and artificial intelligence can bridge the gap.

Artificial intelligence plays an important role to overcome this problem by using different techniques such as food cataloging, using closed-circuit television (CC-TV), laser and machine learning, and sensor-based optical sorting solutions. Artificial Intelligence uses the calibrated machine to manage several sizes of products with less wastage and effective cost. Artificial intelligence-based techniques are very effective for the management of the supply chain. The supply chain is the top priority for all food companies. Machine learning-based technique is useful to improve the supply chain by monitoring and testing product at every step of the supply chain to ensure the compliance of the food industry [43]. Artificial Intelligence based cleaning methods based on ultrasonic sensing and optical fluorescence imaging techniques are being used for the cleaning processes. By using these AI-based techniques cleaning time is reduced by 50 %. These techniques are not only very beneficial but also play a vital role to achieve SDGs' goals by saving water, time and energy.

4.8 Cosmetic artificial intelligence

The automated formulation platforms may enable us to perform complex formulation workflows in a completely automated manner for clean beauty products. The application of artificial intelligence and machine learning plays a significant role in the quest for the development of sustainable personal cake formulations. The modern, technology-driven cosmetic R&D for personal care packaging products are eliminating version plastic components to expendable refillable products to reduce the impact of their product packaging on the planet. We understand very well that with aging and explorer of our skin to ultraviolet light, collagen – the key protein responsible for skin elasticity and structural integrity starts fragmenting. The dermal fibroblast which produces collagen becomes less efficient and results in wrinkles sagging and ununiform pigmentation [44]. Few of the creams such as Tretinoin and Retin-A were used with claims of efficacy but later received comments related to changes in gene expression signatures and other molecular markers. Few skin care companies are still in the phases of developing lotions/creams for enhancing the texture of the skin. Cosmeceuticals are controversial due to the lack of proper definition and limited application in the treatment and prevention of skin conditions by even dermatologists. [45]. In India, we have drug and cosmetic acts/regulations which operate under the authorities of the Central Drug Standard Control Organization (CDSCO). Compliances with good clinical practices and sufficient clinical studies with justified evidence must be available for the demonstration of the intended activity of the cosmeceuticals with details of the concentration of active ingredients and labeling. It's vital to critically understand the potential toxicity and toxicokinetics of the salient chemicals viz parabens, formaldehyde, paraformaldehyde, and additives, etc.

Artificial Intelligence develops several promising utilizes in the field of cosmetic dermatology for the better care of patients. There are a lot of new approaches used in cosmetic industries which are based on artificial intelligence. There are several skin, hair, and personal care products on the shelf of the fast-moving consumable products of the modern market. Currently, most cosmetic companies conduct online surveys and create "quizzes "related to skin and hair care products and cover questions related to patients' demographics. They may include data generated from questionnaires with the help of machine learning. It is anticipated that with the rise of genomic and other associated technologies, skincare products may take measures for improving skin health. Through AI computers thoughtfully assess information about usage or application preferences and interpret logical conclusions for selection criteria. However, as they're machines, they can process a limitless amount of information in fractions of time for ingredients, etc. Many methods such as VISIA skin analysis systems, Life Viz micro by quantifiable, Antera 3D CS from Miravex, and Foto Finder System are in use for skin, hair, and personal care. These all methods depend on artificial intelligence. It is anticipated that models for cosmetic dermatology may help to decide the course of action for skin transplantation or enhance the experiences between dermatologists and patients for improved clinical steps. Artificial intelligence is used in the skin and diagnosis of the skin cancer dataset comprising dermoscopic images of gross lesions. The leaders of dermatologists are finding it necessary to define how the technology of artificial intelligence is turned into clinical practices.

5. Future prospects and conventional challenges

It is anticipated that the global population may increase to another two billion by 2050(FAO) while the arable area may grow by 5 %. Thus, an efficient farming technique is vital to improve the productivity of agricultural products. The repositioning of day-to-day options in aerospace, defense, space exploration, and education is also feasible through the use of AI concepts. Digital transformation and increased operational efficiency for delivering new values added services for customers /stakeholders. In the education sector, the virtual learning environment is becoming more and more user-friendly for personalized learning, intelligent tutoring, universal access to education, surveillance, and evolution.

The cutting-edge breakthroughs in applied science help in material design for sustainability and innovating manufacturing for novel polymers and nanomaterials (using sensors). The AI enables modeling and simulation helps to use specific materials for processes, properties, and performance of material of choices. The non-disrupting inspection, structural health monitoring, and prognostics are the future aspects to be applied. Machine learning has emerged as an innovative predictive model based on mathematical and statistical relationships for several engineering problems [46].

6. Guidelines, rules, and regulations for polymeric manufacturing

In terms of artificial intelligence, WHO has prepared a guidance document for compliance with ethics and governance from a health perspective. We must put the ethics and rights of human civilization at the core of its design, deployments, and usages. The ethical key points are related to issues around reproduction, health care public health for adopting adequate managers for controlling infectious diseases [47,48]. The policymakers and ethicists are planning to draft the updated regulation and policies guidelines, tools to address networking opportunities, and periodically share the new experiences of AI. The frameworks, policies, and best practices for the review of AI-enabled studies are lacking and need strengthening.

Conclusion

Artificial intelligence includes basic, clinical, and public healthy R&D with a health system to assist health services. Clinical screening tools with algorithms for discovery and intervention may help to generate new outputs for human consideration and appropriate action. The case studies may demonstrate good practices and provide insight into unresolved questions. Artificial intelligence and machine learning contain lots of applications over conventional polymeric manufacturing. Artificial intelligence and machine learning technologies have the potential to transform the design and manufacturing the polymers, composites, and nanocomposites with the help of different algorithms and analytical tools. This technology leads to the digitalization of materials design and manufacturing process and challenges researchers and engineers to reconsider and re-evaluate their current technologies to transform traditional methods towards the new era known as smart material and manufacturing for industries 4.0.

Acknowledgment

Authors are thankful to Director, CSIR-IITR, Lucknow and Aligarh Muslim University for encouragement and for providing infrastructural support of the institute for this chapter in the prestigious book of Materials Research Forum LLC, USA

Conflict of Interest

There is no conflict of interest.

Reference

[1] Amisha, P. Malik, M. Pathania, V.K. Rathaur, Overview of artificial intelligence in medicine, J. Family Med Prim Care. 8 (2019) 2328-2331. https://doi.org/10.4103/jfmpc.jfmpc_440_19

[2] What is Artificial Intelligence (AI)? https://www.ibm.com/in-en/cloud/learn/what-is-artificial-intelligence

[3] O. Baloglu, S.Q. Latifi, A. Nazha, What is machine learning? Arch. Dis. Child. Educ. Pract. Ed. 107 (2021) 386-388. https://doi.org/10.1136/archdischild-2020-319415

[4] H. Shi, G. Cao, G. Ma, J. Duan, J. Bai, X. Meng, New progress in artificial intelligence algorithm research based on big data processing of IOT systems on intelligent production lines, Comput. Intell. Neurosci. (2022) 3283165. https://doi.org/10.1155/2022/3283165

[5] S. Bendifallah, A. Puchar, S. Suisse, L. Delbos, M. Poilblanc, P. Descamps, F. Golfier, C. Touboul, Y. Dabi, E. Daarai, Machine learning algorithms as a new screening approach for patients with endometriosis, Sci Rep. 12 (2022) 639. https://doi.org/10.1038/s41598-021-04637-2

[6] S.C. Ligon, R. Liska, J. Stampfl, M. Gurr, R. Mülhaupt, Polymers for 3D printing and customized additive manufacturing, Chem Rev. 117 (2017) 10212-10290. https://doi.org/10.1021/acs.chemrev.7b00074

[7] A. Roda, A.A. Matias, A. Paiva, A.R.C. Duarte, Polymer science and engineering using deep eutectic solvents, Polymers (Basel) 11 (2019) 11050912. https://doi.org/10.3390/polym11050912

[8] M.S.B. Reddy, D. Ponnamma, R. Choudhary, K.K. Sadasivuni, A comparative review of natural and synthetic biopolymer composite scaffolds, Polymers (Basel). 13 (2021) 1105. https://doi.org/10.3390/polym13071105

[9] C.C. Cheng, D.J. Lee, Z.S. Liao, J.J. Huang, Stimuli-responsive single-chain polymeric nanoparticles towards the development of efficient drug delivery systems, Polym. Chem. 7 (2016) 6164-6169. https://doi.org/10.1039/C6PY01623E

[10] Industry E, Industry H. How Industry 4.0 technologies are changing manufacturing.:1 19, 2021.https://onlinelibrary.wiley.com/doi/full/10.1002/bse.2797

[11] Artificial Intelligence: Definition, Types. https://www.britannica.com/technology/artificial intelligence

[12] J. He, S.L. Baxter, J. Xu, X. Zhou, K. Zhang, The practical implementation of artificial intelligence technologies in medicine, Nat Med. 25 (2019) 30-36. https://doi.org/10.1038/s41591-018-0307-0

[13] D.A. Hashimoto, E. Witkowski, L. Gao, O. Meireles, G. Rosman, Artificial Intelligence in Anesthesiology: Current techniques, clinical applications, and limitations, Anesthesiology. 132 (2020) 379-394. https://doi.org/10.1097/ALN.0000000000002960

[14] A.A. Abonamah, M.U. Tariq, S. Shilbayeh, On the commoditization of artificial intelligence, Front. Psychol. 12 (2021) 696346. https://doi.org/10.3389/fpsyg.2021.696346

[15] T. Jiang, J.L. Gradus, A.J. Rosellini, Supervised machine learning: A brief primer, Behav. Ther. 51 (2020) 675-687. https://doi.org/10.1016/j.beth.2020.05.002

[16] J.A. Nichols, H.W.H. Chan, M.A.B. Baker, Machine learning: Applications of artificial intelligence to imaging and diagnosis, Biophys. Rev. 11 (2019) 111-118. https://doi.org/10.1007/s12551-018-0449-9

[17] S.T. Knox, S.J. Parkinson, C.Y.P. Wilding, R.A. Bourne, N.J. Warren, Autonomous polymer synthesis delivered by multi-objective closed-loop optimisation, Polym. Chem. 13 (2022) 1576-1585 . https://doi.org/10.1039/D2PY00040G

[18] N. Patil, iMedPub Journals Polymer Synthesis Characterization and Commercialization. Published online 2018:1-2.https://onlinelibrary.wiley.com/doi/full/10.1002/pola.28976

[19] F. Dong, S. Zhang, J. Zhu, J. Sun, The impact of the integrated development of AI and\ energy industry on regional energy industry: A Case of China, Int. J. Environ. Res. Public Health. 18 (2021) 8946. https://doi.org/10.3390/ijerph18178946

[20] Patil N. iMedPub Journals Polymer Synthesis Characterization and Commercialization. Published online 2018:1-2.https://onlinelibrary.wiley.com/doi/full/10.1002/pola.28976

[21] T. Davenport, R. Kalakota, The potential for artificial intelligence in healthcare, Future Healthc. J. 6 (2019) 94-98. https://doi.org/10.7861/futurehosp.6-2-94

[22] M. Javaid, A. Haleem, R. Vaishya, S. Bahl, R. Suman, A. Vaish, Industry 4.0 technologies and their applications in fighting COVID-19 pandemic, Diabetes Metab. Syndr. 14 (2020) 419-422. https://doi.org/10.1016/j.dsx.2020.04.032

[23] D. Paul, G. Sanap, S. Shenoy, D. Kalyane, K. Kalia, R.K. Tekade, Artificial intelligence in\ drug discovery and development, Drug Discov. Today 26 (2021) 80-93. https://doi.org/10.1016/j.drudis.2020.10.010

[24] K.P. Tran, Artificial intelligence for smart manufacturing: Methods and applications, Sensors (Basel). 16 (2021) 5584. https://doi.org/10.3390/s21165584

[25] M. Mukhtarkhanov, A. Perveen, D. Talamona, Application of stereolithography based 3D printing technology in investment casting, Micromachines (Basel). 11 (2020) 946. https://doi.org/10.3390/mi11100946

[26] S.C. Ligon, R. Liska, J. Stampfl, M. Gurr, R. Mülhaupt, Polymers for 3D printing and] customized additive manufacturing, Chem Rev. 117 (2017) 10212-10290. https://doi.org/10.1021/acs.chemrev.7b00074

[27] L. Jingcheng, V.S. Reddy, W.A.D.M. Jayathilaka, A. Chinnappan, S. Ramakrishna, R. Ghosh, Intelligent polymers, fibers and applications, Polymers (Basel). 13 (2021) 1427. https://doi.org/10.3390/polym13091427

[28] M. Elsabahy, K.L. Wooley, Design of polymeric nanoparticles for biomedical delivery applications, Chem. Soc. Rev. 41 (2012) 2545-2561. https://doi.org/10.1039/c2cs15327k

[29] Z. Guo, J.J. Richardson, B. Kong, K. Liang, Nanobiohybrids: Materials approaches for bioaugmentation, Sci. Adv. 6 (2020) 0330. https://doi.org/10.1126/sciadv.aaz0330

[30] A. Zielińska, F. Carreiró, A.M. Oliveira, A. Neves, B. Pires, D.N. Venkatesh, A. Durazzo, M. Lucarini, P. Eder, A.M. Silva, A. Santini, E.B. Souto, Polymeric nanoparticles: Poduction, characterization, toxicology and ecotoxicology, Molecules 25 (2020) 3731. https://doi.org/10.3390/molecules25163731

[31] E.L. Scheller, P.H. Krebsbach, Gene therapy: Design and prospects for craniofacial regeneration, J. Dent. Res. 88 (2009) 585-596. https://doi.org/10.1177/0022034509337480

[32] S. Shaikh, M. Yaqoob, P. Aggarwal, An overview of biodegradable packaging in the food industry, Curr. Res. Food Sci. 4 (2021) 503-520. https://doi.org/10.1016/j.crfs.2021.07.005

[33] C.M. Carr, D.J. Clarke, A.D.W. Dobson, Microbial polyethylene terephthalate hydrolases:\ Current and future perspectives, Front. Microbiol. 11 (2020) 571265. https://doi.org/10.3389/fmicb.2020.571265

[34] S. Tanaka, K. Wakabayashi, K. Fukushima, S. Yukami, R. Maezawa, Y. Takeda, K. Tatsumi, Y. Ohya, A. Kuzuya, Intelligent, biodegradable, and self-healing hydrogels utilizing DNA quadruplexes, Chem. Asian J. 12 (2017) 2388-2392. https://doi.org/10.1002/asia.201701066

[35] A.E. Stoica, C. Chircov, A.M. Grumezescu, Hydrogel dressings for the treatment of burn wounds: An up-to-date overview, Materials (Basel). 13 (2020) 2853. https://doi.org/10.3390/ma13122853

[36] W. Xin, Y. Gao, B. Yue, Recent advances in multifunctional hydrogels for the treatment of\ osteomyelitis, Front. Bioeng. Biotechnol. 10 (2022) 865250. https://doi.org/10.3389/fbioe.2022.865250

[37] J. Muncke, Tackling the toxics in plastics packaging, PLOS Biol. 19 (2021) 3000961. https://doi.org/10.1371/journal.pbio.3000961

[38] L. Mederake, D. Knoblauch, Shaping EU plastic policies: The role of public health vs. environmental arguments, Int. J. Environ. Res. Public Health. 16 (2019) 3928. https://doi.org/10.3390/ijerph16203928

[39] K.G. Liakos, P. Busato, D. Moshou, S. Pearson, D. Bochtis, Machine learning in agriculture: A review, Sensors (Basel). 18 (2018) 2674. https://doi.org/10.3390/s18082674

[40] S. Nabwire, H.K. Suh, M.S. Kim, I. Baek, B.K. Cho, Review: Application of artificial intelligence in phenomics, Sensors (Basel). 21 (2021) 4363. https://doi.org/10.3390/s21134363

[41] Y.F. Wang, T. Sekine, Y. Takeda, K. Yokosawa, H. Matsui, D. Kumaki, T. Shiba, T. Nishikawa, S. Tokito, Fully printed PEDOT:PSS-based temperature sensor with high humidity stability for wireless healthcare monitoring, Sci. Rep. 10 (2020) 2467. https://doi.org/10.1038/s41598-020-59432-2

[42] N. Ahmed, M.S. Abbasi, F. Zuberi, W. Qamar, M.S.B. Halim, A. Maqsood, M.K. Alam, Artificial intelligence techniques: Analysis, application, and outcome in dentistry-a systematic review, Biomed. Res. Int. (2021) 9751564. https://doi.org/10.1155/2021/9751564

[43] N.R. Mavani, J.M. Ali, S. Othman, M.A. Hussain, H. Hashim, N.A. Rahman, Application of artificial intelligence in the food industry-a guideline, Food Eng. Rev. 14 (2022) 134-175. https://doi.org/10.1007/s12393-021-09290-z

[44] T. Jarvis, D. Thornburg, A.M. Rebecca, C.M. Teven, Artificial intelligence in plastic surgery: Current applications, future directions, and ethical implications, Plast. Reconstr. Surg. Glob. Open. 8 (2020) e3200. https://doi.org/10.1097/GOX.0000000000003200

[45] A.H.A. de Hond, A.M. Leeuwenberg, L. Hooft, I.M.J. Kaant, S.W.J. Nijman, H.J.A.V. Os,\ J.J. Ardoom, T.P.A. Debray, E. Schuit, M.V. Smeden, J.B. Reitsma, E.W. Steyerberg, N.H. Chavennes, K.G.M. Moons, Guidelines and quality criteria for artificial intelligence-based prediction models in healthcare: A scoping review, NPJ Digit. Med. 2 (2022) 1-13. https://doi.org/10.1038/s41746-021-00549-7

[46] L.B. Thomas, S.M. Mastorides, N.A. Viswanadhan, C.E. Jakey, A.A. Borkowski, Artificial intelligence: Review of current and future applications in medicine, Fed. Pract. 38 (2021) 527-538. https://doi.org/10.12788/fp.0174

[47] UNESCO, Artificial intelligence-towards a humanistic approach. https://en.unesco.org/artificial-intelligence

[48] UNESCO, Ethics of Artificial Intelligence. https://www.unesco.org/en/artificial-intelligence/recommendation-ethics

[49] Advanced Polymer Coatings, How AI is influencing the shipping industry today. https://www.adv-polymer.com/blog/artificial-intelligence-in-shipping

[50] Data Flair, Pros and Cons of Artificial Intelligence - A Threat or a Blessing?.https://data-flair.training/blogs/artificial-intelligence-advantages-disadvantages/

[51] TechVidvan, Advantages and Disadvantages of Artificial Intelligence. https://techvidvan.com/tutorials/advantages-and-disadvantages-of-artificial-intelligence/

Chapter 6

Artificial Intelligence for Energy Conversion

Tapasi Ghosh[1*], Bhargavi Koneru[2], and Prasun Banerjee[2]

[1] Ramaiah University of Applied Sciences, Bengaluru, Karnataka, India

[2] Multiferroic and Magnetic Material Research Laboratory (MMMRL), GITAM School of Sciences, Gandhi Institute of Technology and Management (GITAM) University, Bengaluru, Karnataka, India

* tapasi03@gmail.com

Abstract

Many aspects of modern life are dependent on energy of various forms, which has already created strain on natural energy reserves and affected our environment adversely. Scientists and researchers are searching for alternative sources of energy that are sustainable, environment friendly, and renewable. However, any developmental work to invent a material or technique as a new source of energy involves a lengthy and complex experimental process to produce in scale. The last decade has seen remarkable progress in the field of Artificial Intelligence (AI) due to advancements of many new computer hardware, software's, algorithms, technologies, and availability of a large amount of raw input data. We have started harnessing the power of AI to facilitate the process of discovering new materials as alternative energy sources and exploring the different advanced methodologies over the traditional approaches to utilize natural and eco-friendly sources for energy conversion. This book chapter will highlight some of the advancements in Machine Learning and Deep Learning techniques to explore new material resources and methodologies for energy conversion.

Keywords

AI, ML, ANN, Energy Conversion, Catalysts, Microbial Fuel Cell

Contents

Artificial Intelligence for Energy Conversion 123

1. Introduction 124

2. Alternative sources of energy and artificial intelligence 125
3. Machine learning and its application in material sciences 127
4. Limitation of principled method and how ML can intervene 127
5. Application of AI in the domain of energy conversions 128
   5.1   AI in photonics 128
   5.2   AI in electrochemical catalyst 129
   5.3   AI in electrolysis 130
   5.4   AI in fuel cell technology 132
Conclusions 133
Acknowledgments 134
References 134

1. Introduction

Energy is a physical quantity which appears in different forms. Energy cannot be created nor destroyed; instead, it can only be moved from one form to another. During this conversion, the total amount of energy remains conserved. The primary forms of energy stored in objects are either kinetic energy or potential energy. When a ball is thrown upwards, it gains potential energy due to the change in its position and simultaneously falls due to the influence of gravitational force. The potential energy held in the ball is transformed into kinetic energy as a result of the ball's mass and speed. This is a simple example of energy conversion from one form to another. When we rub our palms, the mechanical energy of our body transforms into heat and sound energy. Similarly, a hydroelectric dam turns the kinetic energy of moving water into electric power. Our daily life activities are fulfilled by turning around energies of different forms, and with the ease of modern life, these dependencies are ever-increasing and indispensable.

Though the total amount of energy around us is conserved, we constantly transfer readily available chemical energy (solar electromagnetic energy stored from sunlight) into thermal energy. Energy conversion occurs spontaneously in nature through natural events. However, energy conversion devices initiate the process artificially, e.g., thermocouples convert solar energy into electricity. Solar cells are direct energy conversion devices that produce electricity from optical and electromagnetic radiation. Similarly, other energy conversion devices such as thermoelectric devices and fuel cells are used in power satellites, aerospace systems, power plants, automotive industries, and consumer companies to build a better product for batteries, antennas, etc. Since the average energy consumption of a person has increased in modern life, energy conversion is required to satisfy specific demands of heat or light.

Researchers are working for efficient, effective, reliable, and economically viable energy conversion systems to fulfill the needs of electricity, heating, and cooling for houses, buildings, industrial plants, etc. The R&D work on technological development for various energy conversion systems is ongoing [1] to produce renewable, sustainable, environment-friendly materials and devices, e.g., hydrogen energy conversion systems. Nonrenewable resources (e.g., coal, oil) will deplete quickly over time, and excessive fossil fuel combustion pumps up harmful greenhouse gases to the environment (a potential cause of global warming). On the contrary, the impact of the consumption of renewable energy sources (e.g., sunlight, wind) is benign to the atmosphere. The efficient production of solar and wind energies depends on climate conditions. Whereas safety concerns still undermine the harnessing of nuclear fuel from the controlled fission or fusion process of the nuclear power plant. The discovery of clean energy materials is critical for our modern society's growth.

Artificial Intelligence (AI) has emerged as an essential technology for sustainable, cost-effective renewable energy. AI facilitates communication between air turbines, predicts unscheduled maintenance, reduces costs, regulates power generation flow in smart power plants, and forecasts the weather in advance for solar and wind power plants. For example, Google's Deep Mind [2] is applied for weather forecasts at U.S. wind power facilities to overcome the intermittency challenge. AI is also used to invent new elements and explore critical material properties for energy transformation [3]. AI can accelerate the entire process of discovering new materials through R&D in a laboratory to the commercialization for mass production, which usually takes years.

In this book chapter, we organized the text in the following way: we briefly introduce energy conversion and the mechanism of AI, how models are trained and tested to gain reasoning, which is essential for any Machine Learning algorithm to acquire intelligence. In the next section, we intend to highlight some of the used cases of AI for material science applications such as photonics materials, electrochemical catalysis, electrolysis, and fuel cells technologies, and finally we conclude.

2. Alternative sources of energy and artificial intelligence

Artificial Intelligence is an area of study which allows the computer/machine to acquire the ability to reason artificially, which is in analogy with the cognitive ability of the human brain. It allows a computer system to mimic human cognitive skills such as decision-making and problem-solving. However, the neuron in Artificial Neural Network (ANN) model architecture is a much simplified representation of the structure and synaptic functions of neurons in the human brain. Usually, an AI system is built using machine learning and other techniques. Machine Learning is a sub-branch of AI involving the process through which the computer learns to make decisions without any direct human intervention when trained with large amounts of data. ML is a combination of mathematical logic and model architecture to gain insight into input data, find the hidden pattern, and simultaneously make decisions for unknown and unseen information based upon acquired intelligence.

Frank Rosenblatt designed the deep learning model, among the most basic ANN designs, in 1957 [4]. The Perceptron, also known as the Threshold Logic Unit (TLU), computes a weighted sum of the given inputs, and the output is the outcome of the step function applied over the weighted sum. The training mechanism of Perceptron follows Hebb's Rule, "Cells that fire together, wire together"[5], i.e., the connection between two neurons gets more potent when they fire simultaneously. Modern day ANN refers to the Multilayer Perceptron (MLP) designed by stacking multiple Perceptrons. An MLP architecture usually consists of one input layer, one or more layers of TLUs known as hidden layers, and a final output layer. Until 1986, it was a challenging task to train an MLP, but the invention of back propagation techniques [6] to tune the model parameters automatically has been a breakthrough in the field of AI.

There are multiple reasons for the recent popularity of application of AI in the field of material sciences, e.g., access to open-source machine learning tools and a significant quantum of labeled data to train the algorithms Application Programming Interface (API) resulted in applying the latest developed models and algorithms in a short time. At the same time, recent invention, and advancement of new hardware technologies, e.g., Graphics Processing Unit (GPU) to store and process a large amount of data required to train complex algorithms, is influencing it. The same reasoning is applicable for fostering ML adoption in other research areas as well [7]. ML is replacing some of the traditional research methodologies of inventing new materials or material properties, which were time-consuming involving rigorous processes. In contrast, AI delves into the data, can find the latent representation, and can draw inferences from the unseen data based on the reasoning gained during the training of the models. The Deep Neural Network (DNN) [8] is a model architecture consisting of multiple layers. Each layer holding many neurons can uncover the most relevant features representative of the output to be used in the entire ML job. These characteristics are generated in the first set of layers in the DNN, by providing more weights to the attributes representative of the output. For example, in the case of face recognition tasks, it gives relatively more weightage to the lower-level features related to eye or facial expression. DNN can bypass some of the tedious jobs of feature engineering. Some of the limitations of DNN are the necessity of enough training data for such an extensive network; the training can be prolonged. Generative models, e.g., Generative Adversarial Networks (GAN) depend upon the use of a deep learning process to learn the lower geometrical feature representations from testing phase data in order to predict the result for fresh unseen data based on the learnt distribution. GAN is used to produce realistic simulated data in data-limited situations [9].

3. Machine learning and its application in material sciences

Machine Learning models learn to argue using the following methods: training data, reinforcement classification, moderately supervised learning, and recurrent neural networks [10]. In training data learning, the datasets for which the algorithms are trained are "labeled" data, i.e., a set of output-input pairs. During the data training stage, the algorithm learns the weights, which map the input features to the given output. Most importantly, the model puts more weight on the segments having the most significant impact on the given output. Once the model parameters (weights) are updated, it can make inferences for unseen data. The assumption is much less computation expensive than the training stage. K-Nearest Neighbors, Regression Analysis, Support Vector Machines, and Judgment Trees are examples of key supervised learning algorithms. The training data in unsupervised learning is unlabeled, and the models learn to categories the input data using Clustering, Principal Component Analysis, etc., methods [11].

When developing a Machine learning model, the accessible information is often divided into learning, verification, and testing sets. As explained before, the training dataset is used for training the models to gain insight into the hidden pattern in the data. The validation set helps to choose the best among many trained models. More specifically, one introduces multiple models with various hyper parameters (training rates, variety of hiding layers, neurons in each hidden unit, and so on) and selects the model whose performance is best for the holdout set. Finally, the test set was used to determine how effectively the trained model extrapolates to new scenarios that it has not encountered previously [12].

4. Limitation of principled method and how ML can intervene

Traditionally the electronic structure computational method such as Kohn-Sham density functional theory (DFT) is practiced extensively, which led to the creation of several computational materials databases and subsequently to the discoveries of numerous materials. However, the high predictive accuracy of these methodologies relies on correlation energy functional, which is constructed based on heuristic approaches [13]. ML can fill up this space by enforcing its ability to extract the concise pattern from the trained data and then apply that data-driven knowledge to predict the properties of previously unseen data with minimal computing resources (inference). ML may help to identify complications and create innovative materials with crucial features. For example, a machine learning trained functional on a few molecules can be applied to hundreds of molecules [14]. This is an active research area in the material science community to look for suitable representations of material properties to use the latest ML models [15]. Different material databases, e.g., Resources Group [16] as well as the Experimental Atomic Substances Collection [17], have been designed to cater to the need for a large dataset for training ML models. Even new super hard materials have been discovered by training ML models using computational databases, and other illustrative examples have been reported in [18].

5. Applications of AI in the domain of energy conversions

5.1 AI in photonics

Artificial intelligence has lately been hailed as a critical tool for gaining information and evaluating cause-and-effect links in complicated processes in a massive data environment, notably for operational parameter management and factory automation. When it comes to the application of optoelectronic materials, we find the different aspects of using those materials like sensors, light-absorbing materials, electronic sensors, automatic lights, and LED screens etc., whereas the most and more explored applications come with artificial intelligence, optoelectronic materials can be made from different compounds and compositions [19]. The most produced and used optoelectronic materials in AI are perovskite structured materials. Because of their tailored perovskite structural features, materials such as F, Cl, Br, I, and an alkaline or alkali-earth element are used. J.I. Gomez-Peralta et al. performed a DFT simulation on 136 materials to find their application for AI and better understand their perovskite structure; out of 136 compounds, 96 showed the perovskite compound properties. Artificial neural networking plays a significant role in optoelectronic compounds production. As a result of an analysis process in which they examined the sustainability of the ANN-predicted perovskite structures after experimental and theoretical computations, the molecules predicted by the ANN have been shown to be strong enough to form the perovskite structure. The corner-shared octahedral structure that distinguishes the perovskite structure remained intact after recurrent refinement simulations [20].

ANN has been applied to the maximum number of compounds, and properties have been studied following two different static and dynamic methods. In the static approach, the compounds are synthesized with greater crystal growth and the pros and cons of the compounds are identified as the optimal conditions, temperature dependency, and crystal growth morphologies. With this, the material is created with minimal structural defects [21]. Researchers have derived 81 stimulations from ANN with few inputs and a smaller number of layers. These ANN-derived simulations were used to study the crystal growth structure, the rotational rate of the crystal, temperature dependency of the sample, and the temperature holding capability of the sample containers or crucibles. To describe the complex nonlinear relationship between crystal growth process parameters and interface form, researchers have employed two statistical artificial neural network methods and Gaussian process (G.P.) models [22].

The process involving two statistical approaches was mainly used to identify and optimize magnetic parameters for temperature field control. From this, it is more apparent than in a static method that when both artificial neural networking and Gaussian processes are involved in showing functional group identity and external factor influence, we observe better results compared to ANN alone. Coming to the Dynamic method, it is crucial to inhibit turbulent motion in the melt and manage temperature differences in the crystal, which are responsible for producing harmful crystal defects and unwanted fluctuation in crystal diameter [23].

The dynamic ANN method has been applied to overcome these crystal diffusions, and crystal structure studies were made in the static feed dynamic crystal growth approach, the researchers evaluated the process parameters of two warmers as well as the velocities of the thermal barrier throughout the particles generated. Precise temperature control is required for the Metal-Organic Chemical Vapor Deposition (MOCVD) of GaN for micro - scale and absorption spectra. Preserve wavelength homogeneity and manage wafer bow and decrease wafer slide. NAXR neural networks have proven successful for numerous time-series modeling jobs, notably in control applications, by learning long-term data analysis applications using open-source massive crystallization process data will eliminate the final barrier for ANN applications and considerably expedite the development of new novel crystal substance products. [24]

5.2 AI in electrochemical catalyst

On the basis of evidence of the start spinning DFT computational and dynamic modeling, comprehensive knowledge of an implementation of empirically achievable single-atom electrochemical catalyst for H_2O_2 creation has been described. When the electrode potential reaches equilibrium, the $\Delta g°$ changes in each fundamental step are equal, permitting all interaction free energies to be zero. The Gibbs energy diagram of oxygen gas conversion to H_2O_2, where the computer initially evaluated the sustainability and selectivity of 31 single-atom catalysts, is one of the examples [25]. By considering the seven single-atom catalysts with different combinations of single crystal atoms, the DFT simulations are to make under the effect on H_2O_2 and then identify six among the top 7 highly potential catalysts have macro cyclic structures, implying that in this study, macro cyclic structures surpass graphene and other materials [26]. Metals having a weaker oxygen affinity, such as Ag, Au, and Pd, can significantly reduce band recombination between the metal and oxygen, resulting in greater preference favoring H_2O_2 production. This explains how such materials are widely utilized in ORR full form? as electro catalysts with a two-electron O2 reduction ratio [27].

Analyzing a single-atom catalyst (SAC) and utilizing Machine learning techniques to investigate the affinity and specificity of a single atom catalyst the proportional connections between the Gibbs free energy and the proportionality connections in between the (OOH) and carboxylic group illustrate the variations in G (O) and G (OOH) on the 31 studied SACs ML explains why SAC selectivity and functionality differ and how the development of a more effective and reliable SAC for hydrogen peroxide generation. As can be shown, ML may significantly aid in establishing the link between material structure and attributes [28]. The technique was utilized to pick distinctive variables and provided the catalytic efficiency prediction equation. Combining two separate transverse magnetization decay curves (TMDCs) while altering the rotary orientation, bonding duration, spacing among layers, and band gap proportion of two different materials, performance may be significantly improved. In terms of creating a very sophisticated and highly specialized electro catalytic activity for oxygen reductions and release, researchers have summarized the process of developing precise descriptors and discovered different types of Descriptive tags that can increase the capacity to forecast material characteristics by applying machine learning and increased transmission processing to produce new as well as other catalysts compounds.[29].

5.3   AI in electrolysis

The most efficient technique for producing hydrogen is water electrolysis. However, the cathode materials are primarily costly metals that are inappropriate for large-scale usage. In such situations, computational methods play a role in identifying the different techniques for producing hydrogen-like materials [30]. Support Vector Machine methods and Artificial Neural Networks were used to explore the atomic characteristics, which demonstrated a significant relationship among both the catalytic site and bond angles parameters at the atomic level [31]. Changes beyond the original input set's prediction may indeed be attributed to a reduced amount of low performance samples in the original database. SVR's effectiveness is confirmed by the defined metrics. Over-fitting of all algorithms is possible, although it is dependent on the quantum of data used for training. As a result, as is customary, the facts determine the forecast rather than the other way around. To find a superior neural network approach, different numbers of neurons in the hidden layer were placed behind the four input neurons and in front of the one-bit output neuron [32]. Neuron densities in a simple linear plane range from 2 to 30 were tested, with modest networks with hidden layers of 6 or 8 neurons surpassing the rest. Decreased neuron quantities resulted in far more inaccuracy, but increased neuron quantities risked overfitting the network.[33]. Adding extra dropout layers to avoid overfitting had little impact. The network was trained across 500 epochs to avoid overfitting, with the option of quitting early if the metrics did not change. For validation, a 20% partition of something like the training dataset was used. Scientists picked a 46661 configuration of neurons and functioned as an input signal for this confirmation, as seen in the Fig: 1 [34].

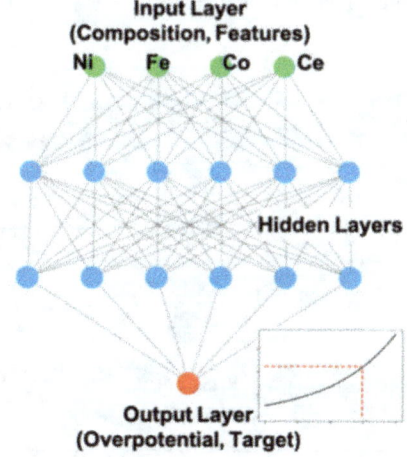

Fig: 1. ANN input and output nodes [34]

The combined use of artificial neural networks, other types of AI technologies, and the Machine learning approach to assess performance and forecast atomic qualities projected more stable crystal structures. The researchers then trained and predicted the electro-chemical data set using three different ML models, organizing, and modifying the data, overseeing the learning process, and assessing the three strategies using test data. Their empirical investigation shows that use of highly complex patterns does not necessarily result in more accurate prediction results, and that model accuracy is linked to the data itself [34].

5.4 AI in fuel cell technology

Although computational equations could be used to characterize the Microbial Fuel Cells (M.F.C.) biofilm formation mechanism and the physicochemical consequences, they cannot attain the desired results when simulating complex settings such as varied and combined microbial populations. To create mathematical correlations between wastewater/solution parameters, biofilm populations, and reactor efficiency, Artificial Neural Networks (ANNs) were applied. ANN simulations that included biotic interactions projected reactor efficiency better than those that did not. Because of the intricate interactions that take place in mixed species in bio electrochemical reactors such as microbial fuel cells (MFCs), as shown in Fig. 2, it is challenging to anticipate performance outcomes under experimental situations.

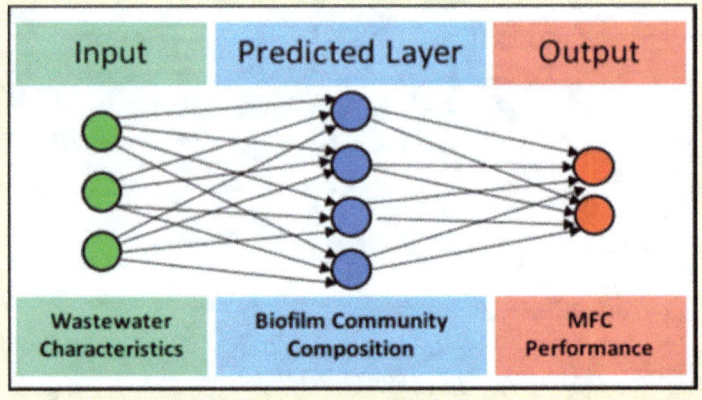

Fig: 2. MFCs interaction in the mixed-phase system [35]

It was challenging to construct thorough probabilistic models for different and dynamic microbial communities; nevertheless, at this time, the ANN-based simulations made it simple to reliably assess a variety of reactor output and biofilm compositions throughout water treatment. The algorithms that link community composition, wastewater parameters, and reactor function have the ability to find the previously unknown and unverified waste waters that result in high MFC productivity. The effectiveness of data mining techniques is determined by the integrity of the data set. Several factors, including reactor architecture, operating parameters, and starting biofilm communities, were regulated in the current study to ensure that results correctly reflected connections between the parameter estimates [35].

Table 1: Simulation Process

Domain of Energy Conversions	First Generation	Second Generation	Third Generation
AI in Photonics	Structural properties calculation	Crystal structure prediction	Statically driven design
AI in electrochemical Catalyst	Local optimization algorithm	Global optimization algorithm	Chemical and physical data computation
AI in Electrolysis	Data collection	Experimental analysis	Simulated data comparisons
AI in Fuel Cell Technology	ANN Model Development	Multivariate Polynomial Regression (MPR) Method	Comparison of the ANN and M-ANN [36, 37]

Conclusions

AI is playing a pivotal role in empowering sustainable and economic forms of power generation. The energy sector has already deployed Al and machine learning for various functionalities, e.g., smart grids to balance electricity demand and supply, sensor-connected power plants, and windmills to mitigate the effects of low airflow in some regions. Still, wind and solar renewable energy suffer from intermittency problems. Catalyst is a crucial ingredient in sustainable energy conversion systems. The invention of suitable materials for catalysts is time-consuming, requiring theoretical domain knowledge and a rigorous experimental process. Deploying different modern machine learning algorithms and architectures and combining different hyper-parameters and optimization techniques, researchers can discover several catalysts for energy conversion systems. ML algorithms can virtually scan the existing databases and identify materials' novel characteristics and properties using a data-driven approach. It compensates for the said limitations of the current conceptual framework. It can immediately analyze and anticipate material properties and crystal structures, as well as build new materials. AI can overcome human shortcomings in temporal analysis and processing. Secondly, AI can bring the reduction in time necessary for commercialization by reducing the material development cycle. Previously, to adjust and improve experimental circumstances, researchers had to read hundreds of thousands of relevant publications. They can only rely on human classification and summarization to select items for required characteristics, which are significant and time consuming. Using AI techniques increase efficiency through modeling and optimization without raising costs can be achieved. Finally, AI shows promise in material prediction. Nonetheless, there are several obstacles to overcome.

Acknowledgments

T. Ghosh acknowledges the DST Grant WOS-A/PM-42/2018 to support the research activities. PM P. Banerjee thanks SERB, India, for a TAR/2021/000032 research grant. B Koneru thanks GITAM University for Dr M. V.V. S Murthi research fellowship.

References

[1] I. Dincer, Comprehensive Energy Systems, first ed., Elsevier, 2018.

[2] J. Powles, H. Hodson, Google DeepMind and healthcare in an age of algorithms, Health Technol. (Berl). 7 (2017) 351-367. https://doi.org/10.1007/s12553-017-0179-1

[3] X. Yang, Z. Luo, Z. Huang, Y. Zhao, Z. Xue, Y. Wang, W. Liu, S. Liu, H. Zhang, K. Xu, Development status and prospects of artificial intelligence in the field of energy conversion materials, Front. Energy Res. 8 (2020) 167. https://doi.org/10.3389/fenrg.2020.00167

[4] F. Rosenblatt, The Perceptron: A perceiving and recognizing automaton, Cornell University, Ithaca, NY, Project PARA, Cornell Aeronautical Laboratory, Rep. (1957) 85-460.

[5] D.O. Hebb, The first stage of perception: Growth of the assembly, in: J.A. Anderson, E. Rosenfeld (Eds.), The Organization of Behavior, Wiley, New York, 1949, pp. 60-78.

[6] R.L. Watrous, L. Shastri, A.H. Waibel, Learned phonetic discrimination using connectionist networks, Proc. European Conference on Speech Technology. 1987, 1377-1380.

[7] K. Naidu, N.S. Kumar, P. Banerjee, B. Reddy, A review on the origin of nanofibers/nanorods structures and applications, J. Mater. Sci. - Mater. Med. 32 (2021) 1-25. https://doi.org/10.1007/s10856-021-06541-7

[8] Y. Bengio, Y.L. Cun, Scaling learning algorithms towards AI, in: L. Bottou, O. Chapelle, D. Decoste, J. Weston (Eds.), Large-Scale Kernel Machines, MIT Press, 2007, pp. 1- 41.

[9] K.R.M. Rao, K. Haripriya, P. Banerjee, A. Franco, Microbiologically influenced corrosion, in: N.S. Kumar, P. Banerjee, H. Manjunatha, K.C.B. Naidu (Eds.), Corrosion Science: Modern Trends and Applications, Bentham Science, 2021, pp 121-146. https://doi.org/10.2174/9789811481833121010011

[10] A. Géron, Hands-on Machine Learning with Scikit-Learn, Keras, and TensorFlow: Concepts, Tools, and Techniques to Build Intelligent Systems, O'Reilly Media, Inc., 2019.

[11] N.S. Kumar, K.C.B. Naidu, P. Banerjee, T.A. Babu, B.V.S. Reddy, A review on metamaterials for device applications, Crystals (Basel). 11 (2021) 518. https://doi.org/10.3390/cryst11050518

[12] M.R. Jedla, B. Koneru, A.F. Jr, D. Rangappa, P. Banerjee, Recent developments in nanomaterials based adsorbents for water purification techniques, Biointerface Res. Appl. Chem. 12 (2021) 5821-5835. https://doi.org/10.33263/BRIAC125.58215835

[13] Q. Yan, J. Yu, S.K. Suram, L. Zhou, A. Shinde, P.F. Newhouse, W. Chen, G. Li, K.A. Persson, J.M. Gregoire, Solar fuels photoanode materials discovery by integrating high-throughput theory and experiment, Proc. Natl. Acad. Sci. 114 (2017) 3040-3043. https://doi.org/10.1073/pnas.1619940114

[14] R. Nagai, R. Akashi, O. Sugino, Completing density functional theory by machine learning hidden messages from molecules, NPJ Comput. Mater. 6 (2020) 1-8. https://doi.org/10.1038/s41524-020-0310-0

[15] G.R. Schleder, A.C.M. Padilha, C.M. Acosta, M. Costa, A. Fazzio, From DFT to machine learning: Recent approaches to materials science-A review, J. Phys.: Mater. 2 (2019) 032001. https://doi.org/10.1088/2515-7639/ab084b

[16] A. Jain, S.P. Ong, G. Hautier, W. Chen, W.D. Richards, S. Dacek, S. Cholia, D. Gunter, D. Skinner, G. Ceder, Commentary: The Materials Project: A materials genome approach to accelerating materials innovation, APL Mater. 1 (2013) 011002. https://doi.org/10.1063/1.4812323

[17] S. Kirklin, J.E. Saal, B. Meredig, A. Thompson, J.W. Doak, M. Aykol, S. Rühl, C. Wolverton, The Open Quantum Materials Database (OQMD): aAssessing the accuracy of DFT formation energies, NPJ Comput. Mater. 1 (2015) 1-15. https://doi.org/10.1038/npjcompumats.2015.10

[18] C.P. Gomes, B. Selman, J.M. Gregoire, Artificial intelligence for materials discovery, MRS Bulletin. 44 (2019) 538-544. https://doi.org/10.1557/mrs.2019.158

[19] P. Banerjee, A. Franco, K.C.B. Naidu, N.S. Kumar, Water-borne polyurethane metal oxide nanocomposite applications, in: Inamuddin, R. Boddula, A. Khan (Eds.), Sustainable Production and Applications of Waterborne Polyurethanes, Springer, 2021, pp. 155-169. https://doi.org/10.1007/978-3-030-72869-4_10

[20] J.I.G. Peralta, X. Bokhimi, Ternary halide perovskites for possible optoelectronic applications revealed by Artificial Intelligence and DFT calculations, Mater. Chem. Phys. 267 (2021) 124710. https://doi.org/10.1016/j.matchemphys.2021.124710

[21] P. Banerjee, A. Franco Jr, K.C.B. Naidu, A. Khan, A.M. Asiri, S. Natarajan, Metal-organic framework-based materials and renewable energy, in: A. Khan, F. Verpoort, A.M. Asiri, M.E. Hoque, A. Bilgrami, M. Azam, K.C.B. Naidu (Eds.), Metal-Organic Frameworks for Chemical Reactions, Elsevier, 2021, pp. 153-166. https://doi.org/10.1016/B978-0-12-822099-3.00008-3

[22] B. Koneru, J. Swapnalin, S. Natarajan, A. Franco Jr, P. Banerjee, Intercalation of nanoscale multiferroic spacers between the two-dimensional interlayers of MXene, ACS Omega.7 (2022) 20369-20375. https://doi.org/10.1021/acsomega.2c02471

[23] P. Banerjee, A.F. Jr, R.Z. Xiao, Effects of Y and Ni co-doping in $Bi_2Fe_4O_9$ $BiFeO_3$ based multiferroic ceramics, Mater. Today: Proc. 46 (2021) 4716-4719. https://doi.org/10.1016/j.matpr.2020.10.302

[24] N. Dropka, M. Holena, Application of artificial neural networks in crystal growth of electronic and opto-electronic materials, Crystals (Basel) 10 (2020) 663. https://doi.org/10.3390/cryst10080663

[25] P.V.V. Romanholo, T.E.P. Alves, J. Swapnalin, P. Banerjee, A.F. Jr, Tailoring the magnetic properties of Zn doped Nickel, Magnesium and Cobalt Ferrite ceramics, Mater. Chem. Phys. 284 (2022) 126072. https://doi.org/10.1016/j.matchemphys.2022.126072

[26] K.C.B. Naidu, N.S. Kumar, R. Boddula, S. Ramesh, R. Pothu, P. Banerjee, M. Sarma, H. Manjunatha, B. Kishore, Recent advances in nanomaterials for Li-ion

batteries, in: Inamuddin, R. Boddula, M.F. Ahmer, A.M. Asiri (Eds.), Lithium-Ion Batteries: Materials and Applications, Materials Research Forum, 2020, pp. 148 160.

[27] X. Guo, S. Lin, J. Gu, S. Zhang, Z. Chen, S. Huang, Simultaneously achieving high activity and selectivity toward two-electron O2 electroreduction: The power of single-atom catalysts, ACS Catal. 9 (2019) 11042-11054. https://doi.org/10.1021/acscatal.9b02778

[28] B. Koneru, J. Swapnalin, P. Banerjee, K.C.B. Naidu, N.S. Kumar, Materials under extreme pressure: Combining theoretical and experimental techniques, Eur. Phys. J. Spec. Top. 137 (2022) 1-12.

[29] X. Yang, Z. Luo, Z. Huang, Y. Zhao, Z. Xue, Y. Wang, W. Liu, S. Liu, H. Zhang, K. Xu, Development status and prospects of artificial intelligence in the field of energy conversion materials, Front. Energy Res. 8 (2020) 167. https://doi.org/10.3389/fenrg.2020.00167

[30] M. Prakash, N.S. Kumar, K.C.B. Naidu, M. Sarma, P. Banerjee, R.J. Kumar, R. Pothu, R. Boddula, Electrode materials for K-ion batteries and applications, in: Inamuddin, R. Boddula, A.M. Asiri (Eds.), Potassium-Ion Batteries: Materials and Applications, Wiley, 2020, pp.123-136. https://doi.org/10.1002/9781119663287.ch5

[31] J. Hu, X. Cao, X. Zhao, W. Chen, G. Lu, Y. Dan, Z. Chen, Catalytically active sites on Ni5P4 for efficient Hydrogen evolution reaction from atomic scale calculation, Front. Chem. 7 (2019) 444. https://doi.org/10.3389/fchem.2019.00444

[32] R.S. Melo, A.F. Jr, P. Banerjee, Nanoscale-driven single-domain structure in Nickel substituted superparamagnetic Cobalt Ferrites, Solid State Commun. 341 (2022) 114560. https://doi.org/10.1016/j.ssc.2021.114560

[33] P. Banerjee, A.F. Jr, R.Z. Xiao, K.C.B. Naidu, R.M. Rao, R. Pothu, R. Boddula, Advancement in electrolytes for rechargeable batteries, in: R. Boddula, Inamuddin, R. Pothu, A.M. Asiri (Eds.), Rechargeable Batteries: History, Progress, and Applications, Wiley, 2020, pp. 87-98. https://doi.org/10.1002/9781119714774.ch5

[34] R. Palkovits, S. Palkovits, Using artificial intelligence to forecast water oxidation catalysts, ACS Catal. 9 (2019) 8383-8387. https://doi.org/10.1021/acscatal.9b01985

[35] K.L. Lesnik, H. Liu, Predicting microbial fuel cell biofilm communities and bioreactor performance using artificial neural networks, Environ. Sci Technol. 51 (2017) 10881-10892. https://doi.org/10.1021/acs.est.7b01413

[36] Z. Yang, B. Wang, X. Sheng, Y. Wang, Q. Ren, S. He, J. Xuan, K. Jiao, An artificial intelligence solution for predicting short-term degradation behaviors of proton

exchange membrane fuel cells, Appl. Sci. 11 (2021) 6348. https://doi.org/10.3390/app11146348

[37] X. Yang, Z. Luo, Z. Huang, Y. Zhao, Z. Xue, Y. Wang, W. Liu, S. Liu, H. Zhang, K. Xu, Development status and prospects of artificial intelligence in the field of energy conversion materials, Front. Energy Res. 8 (2020) 167. https://doi.org/10.3389/fenrg.2020.00167

Keyword Index

AI Tools	87
ANN	123
Catalysts	123
Composites	1
Databases	87
Deep Learning	24,47
Energy Conversion	123
Innovation	105
Intelligence	105
Internet of Things	47
Material Genomics	87
Microbial Fuel Cell	123
Nanorobots	1
Nanotechnology	1
Polymer	105
Renewable Energy	47
Robotics	1
Solar Energy	47
Solar Photovoltaic	47
Sustainable	105

About the Editors

Dr. Inamuddin is working as Assistant Professor at the Department of Applied Chemistry, Aligarh Muslim University, Aligarh, India. He obtained a Master of Science degree in Organic Chemistry from Chaudhary Charan Singh (CCS) University, Meerut, India, in 2002. He received his Master of Philosophy and Doctor of Philosophy degrees in Applied Chemistry from Aligarh Muslim University (AMU), India, in 2004 and 2007, respectively. He has extensive research experience in multidisciplinary fields of Analytical Chemistry, Materials Chemistry, and Electrochemistry and, more specifically, Renewable Energy and Environment. He has worked on different research projects as project fellow and senior research fellow funded by University Grants Commission (UGC), Government of India, and Council of Scientific and Industrial Research (CSIR), Government of India. He has received Fast Track Young Scientist Award from the Department of Science and Technology, India, to work in the area of bending actuators and artificial muscles. He has also received the Sir Syed Young Researcher of the Year Award 2020 from Aligarh Muslim University. He has completed four major research projects sanctioned by University Grant Commission, Department of Science and Technology, Council of Scientific and Industrial Research, and Council of Science and Technology, India. He has published 207 research articles in international journals of repute and nineteen book chapters in knowledge-based book editions published by renowned international publishers. He has published 165 edited books with Springer (U.K.), Elsevier, Nova Science Publishers, Inc. (U.S.A.), CRC Press Taylor & Francis Asia Pacific, Trans Tech Publications Ltd. (Switzerland), IntechOpen Limited (U.K.), Wiley-Scrivener, (U.S.A.) and Materials Research Forum LLC (U.S.A). He is a member of various journals' editorial boards. He is also serving as Associate Editor for journals (Environmental Chemistry Letter, Applied Water Science and Euro-Mediterranean Journal for Environmental Integration, Springer-Nature), Frontiers Section Editor (Current Analytical Chemistry, Bentham Science Publishers), Editorial Board Member (Scientific Reports-Nature) and Review Editor (Frontiers in Chemistry, Frontiers, U.K.) He has also guest-edited various special thematic special issues to the journals of Elsevier, Bentham Science Publishers, and John Wiley & Sons, Inc. He has attended as well as chaired sessions in various international and national conferences. He has worked as a Postdoctoral Fellow, leading a research team at the Creative Research Initiative Center for Bio-Artificial Muscle, Hanyang University, South Korea, in the field of renewable energy, especially biofuel cells. He has also worked as a Postdoctoral Fellow at the Center of Research Excellence in Renewable Energy, King Fahd University of Petroleum and Minerals, Saudi Arabia, in the field of polymer electrolyte membrane fuel cells and computational fluid dynamics of polymer electrolyte membrane fuel cells. He is a life member of the Journal of the Indian Chemical Society. His research interest includes ion exchange materials, a sensor for heavy metal ions, biofuel cells, supercapacitors and bending actuators.

Ms. Maha Khan is a Research Scholar at the Department of Applied Chemistry, Aligarh Muslim University (A.M.U.), Aligarh, India. She has also pursued her Bachelor's in Chemistry and Master's in Polymer Science and Technology from A.M.U., Aligarh. Her research work focuses primarily on Enzymatic Biofuel Cells, a pathway to clean and green energy.

Dr. Mohammad A. Jafar Mazumder has been serving as a Professor of Chemistry at King Fahd University of Petroleum & Minerals (KFUPM), Saudi Arabia. He has extensive experience in designing, synthesizing, and characterizing various organic compounds, ionic and thermo-responsive polymers for corrosion, water treatment, and biomedical applications. Dr. Jafar Mazumder obtained his B.Sc (Hons.), M.Sc (Chemistry) from Aligarh Muslim University, India, MS (Chemistry) from KFUPM, Saudi Arabia, and Ph.D. in Chemistry (2009) from McMaster University, Canada.

In more than 20 years of academic research, Dr. Jafar Mazumder has had the opportunity to work with several international collaborative research groups and has exposed himself to a broad range of research areas. Dr. Jafar Mazumder secured 7 US patents, published more than 85 articles in peer-reviewed journals, 37 conference proceedings, 9 book chapters, and co-edited 4 books with Springers and Trans Tech publications. He is awarded as a Fellow of the Royal Society of Chemistry and Chartered Chemist, Association of Chemical Profession of Ontario, Canada. Besides, Dr. Jafar Mazumder is a member of the American Chemical Society (ACS), Canadian Society for Chemistry (CSC), Canadian Biomaterial Society (CBS), and a life member of Bangladesh Chemical Society (BCS). In his academic career, he was awarded numerous national and international scholarships and awards that include the prestigious Indian Council for Cultural Relations (ICCR) Scholarship from Govt. of India for undergraduate studies in India, Aligarh Muslim University undergraduate & graduate Gold medal, and certificate of excellence from the Ministry of Human Resource Development, Govt. of India, and MITACS postdoctoral fellowship (Canada) for pursuing postdoctoral research in Chemical and Biomedical Engineering.

Currently, Dr. Jafar Mazumder is actively involved in several ongoing university (KFUPM), government (KACST, NSTIP), and client (Saudi Aramco) funded projects in the capacity of principal and co-investigators. His current research interest includes the design, synthesis, and characterization of various modified monomers and polymers for potential use in the inhibition of mild steel corrosion in oil and gas industries and the preparation of multilayered polyelectrolyte coated membranes for the removal of heavy metals and organic contaminants from aqueous water samples. Long term scientific goal of Dr. Jafar Mazumder is not merely to make science fun and entertaining for people. It is to engage them with a multidisciplinary scientific mission at a deeper level to create a space through which they can interact with scientific ideas, develop connections between science, engineering, and biology, and thoughts of their own to contributions to society. He feels this goal and engaging personality make him a pleasant person to work with and help inspire his co-workers in any professional setting.